SCIENCE SOCIETIES

Resources for Life in a Technoscientific World

Sarah R. Davies

First published in Great Britain in 2025 by

Bristol University Press
University of Bristol
1–9 Old Park Hill
Bristol
BS2 8BB
UK
t: +44 (0)117 374 6645
e: bup-info@bristol.ac.uk

Details of international sales and distribution partners are available at bristoluniversitypress.co.uk

British Library Cataloguing in Publication Data
A catalogue record for this book is available from the British Library

ISBN 978-1-5292-2899-1 hardcover
ISBN 978-1-5292-2900-4 paperback
ISBN 978-1-5292-2901-1 ePub
ISBN 978-1-5292-2902-8 ePdf

Cover design: Nicky Borowiec
Front cover image: Stocksy/Audshule
Bristol University Press uses environmentally responsible print partners.
Printed and bound in Great Britain by CPI Group (UK) Ltd, Croydon, CR0 4YY

FSC
www.fsc.org
MIX
Paper | Supporting responsible forestry
FSC® C013604

Contents

List of Boxes

Acknowledgements

It's a rainy day in Aachen, Germany, and I really should be finishing the book manuscript you are about to read, but instead I have decided to warm myself up, at least figuratively, by writing the acknowledgements. It gives my heart joy to think of all the people who have helped me in this project, and who have been so generous with their time, suggestions, and support.

This book is dedicated to all of the students I have been fortunate enough to engage with, particularly those on the Masters programme 'Science – Technology – Society' run by my home department of Science and Technology Studies at the University of Vienna. I have learned so much from our smart, engaged, motivated, and passionate community of students, and am deeply grateful for their thoughtful engagement with the topics that I write about in this book. Of that community, particular thanks go to Elaine Goldberg, Nora Ederer, and Constantin Holmer, all of whom worked as research assistants with me while I was working on this project and who helped me with researching cases and examples for it. Without Nora, especially, this book would be much shorter, drier, and even less diverse, while Elaine was an insightful and encouraging first reader of it. I am hugely grateful for their assistance and want to acknowledge their work: I truly couldn't have completed this text without them. I'd also like to thank Cyntha Wieringa, Harry Wynands, Anastasia Nesbitt, Dario Feliciangeli, and Tereza Butková, all of whom have worked as teaching assistants during my time at the department and who have offered all kinds of intellectual, practical, and moral support. It has been a privilege to work with them, and indeed – again – with our student cohorts as a whole. Thank you.

I am also embedded in a community of colleagues at Vienna who constantly help me think about what Science and Technology Studies is and what it should be. Ulrike Felt and Max Fochler shaped the programme on which I now teach, and I never fail to gain insights from their scholarship, while Nina Klimburg-Witjes is similarly a source of Science and Technology Studies wisdom. Karin Neumann, Ulrike Ritter, Fabiola Klaus, and Katrin Hackl keep things running, while Andrea Schikowitz, Ariadne Avkıran, Bao-Chau Pham, Esther Dessewffy, Fredy Mora Gámez, Kathleen Gregory, and Rafaela Cavalcanti de Alcantara are my closest collaborators, whose work supports mine in multiple ways. Thank you, thank you, thank you.

The book has also benefited from a network beyond Vienna, both in terms of comments and suggestions on it and the places in which it was written and discussed. Barbara Prainsack, Emily Dawson, Joseph Roche, Simon Lock, Raffael Himmelsbach, and Michael Penkler were kind enough to read drafts of the manuscript and to offer hugely helpful feedback on it, as did one extremely generous anonymous reader via Bristol University Press. Again, without those suggestions the book would be much weaker (though I take responsibility for the places where I failed to fully implement that feedback). I have presented or discussed the book at Copenhagen's Medical Museion and at the Technical University of Denmark, and want to thank Maja Horst, Monamie Bhadra Haines, Louise Whiteley, and Cecilie Glerup for the opportunity to do so and for the rich conversations I have had with each of them. In addition the book (and my energy for writing it) has benefitted from visits to the New Institute Centre for Environmental Humanities at Ca' Foscari University of Venice, Venice International University, and the Käte Hamburger Kolleg Aachen (from where I am currently writing – I don't blame them for the rain). Thank you in particular to Francesca Tarocco, Alessandra Fornetti, Ilda Mannino, Elisa Carlotto, Stefan Böschen, and Gabriele Gramelsberger for their generosity in hosting me, and for the insights I have gained from them.

There is also a huge network of people who have influenced this book without knowing it: the friends, colleagues, and collaborators I have worked and interacted with over the years and learned so much from. Jane Calvert, Rob Smith, Alan Irwin, Matt Kearnes, Phil Macnaghten, Cynthia Selin, Erik Aarden, Barbara Streicher, Susanne Öchsner, Jane Gregory, Felicity Mellor, Noriko Hara, Karen Kastenhofer, Katja Mayer, Doro Born, Erika Szymanski, Nina Amelung, Karin Tybjerg, Bruce Lewenstein, Megan Halpern, Katrine Lindvig, Sophie Lecheler, Torsten Möller, Laura Kösten, Rebecca Wells, Fabiana Zollo, Ellie Armstrong, Sophie Gerber, Sarah Schönbauer, Tereza Stöckelová, Rasmus Kvaal Wardemann, Pouya Sepehr, Virginia Vargolskaia, Lisa Sigl, Florentine Frantz, Sonja Jerak-Zuiderent, Melanie Smallman, Alice Bell, Rhian Salmon, Stevie de Saille, Warren Pearce, Dave Guston, and Fares Kayali are just some of these people. I have also learned a lot from the Masters students whose theses I have supervised in recent years, and in particular from Livia Beck, Jana Wiese, Fabian Saxinger, Keziah Sheldon, Celine Rabé, Hanna Hämäläinen, Luise Lehfeldt, Kathrin Runggatscher, and Auriane Sabine van der Vaeren. I'd also like to thank Paul Stevens and Ellen Mitchell at Bristol University Press for their support and encouragement, even in the face of my ever-optimistic timetabling and inevitable failure to meet deadlines, Gail Welsh and Kelly Winter for their generous and patient assistance with copy editing, and Ally Davies, who helped manage the chaos that is my referencing system.

To the Bredon girls: Emma, Amy, Buffy, Helen, Rachel, and Jess. I think of you often, while I have been writing this manuscript as well as more generally. I'm just sorry that this is so much more vanilla than Emma's offering.

So much kindness, so much wisdom. I will close by thanking my partner, Raffael Himmelsbach, without whom I would be stupider, sadder, and hungrier. I owe him so much: he has acted as reader and commenter, sounding board, book-hunter, and ethical compass. Thank you.

1

Introduction: Science Societies

I am writing from what feels like a time of crisis. As I sit in my office in Vienna the COVID-19 pandemic continues to rage, causing everything from cancelled meetings and online teaching to millions of excess deaths across the world. The climate crisis – the onset of changes in the global climate caused by human activity – is beginning to shape weather patterns and the degree to which particular regions of the world are habitable; each year, we see more extreme weather, alongside disasters such as widespread flooding or wildfires. And there are conflicts and clashes at national borders. Two countries over from where I sit in Austria, Russia has invaded Ukraine, and the country is the site of appalling violence as it fights to maintain its sovereignty. This is, however, just one example of forms of nationalistic aggression that are taking place around the world, from Colombia to Afghanistan, which are causing widespread death, destruction, and displacement. Many predict that such conflicts will only increase as climate change reshapes the world's landscapes.

This is not a book about these crises, or the many others that shape our world. It is, however, a book about one thing that these events have in common. In all of these examples, scientific and technical knowledge and expertise are central to how they are understood, managed, and unfold. While they are not *only* scientific crises or controversies, science and technology are vital aspects of them. To take some examples: I have just read an expert commentary on the war in Ukraine that uses the results from 'war games' to discuss possible outcomes of the current situation.[1] These highly technical processes use modelling to try and understand different conflict scenarios, with the results of such games themselves feeding in to political advice and decision making. I have also just carried out a PCR test for COVID-19, a now regular occurrence to check whether I am infected and whether I can safely meet with others. My results will come back in 24 hours: I have become adept both at carrying out the test and reading the results (I have learned what a 'CT value' is, for instance).[2] And my newsfeed continues to be filled with debate and discussion of different technical aspects of the climate crisis: are carbon capture technologies viable for mitigating climate change?[3] Exactly which lifestyle changes will have the most impact?[4] What effects will different average temperature rises have?[5]

***

As the examples suggest, I wrote these opening paragraphs early on in the process of writing this book, at the start of 2022. While developing the Introduction before the other chapters are finished goes against all book-writing advice (and while the book has certainly changed a lot over the months I have been working on it), I still think they capture something central to my argument. *We live in a technoscientific world.* Across the globe, science and technology are key aspects of public policy and advice, everyday experience, and popular culture. Our days are marked by mundane technologies (from mobile phones to traffic lights) and shaped by expert advice (as we decide what to eat, study, or grow for food). We consume science by watching science channels on YouTube or by researching the medical conditions that affect our lives. Science and technology – captured by the portmanteau word *technoscience* – are, in different ways, central to life in the vast majority of contemporary societies around the world. They affect us on both individual and collective levels, shaping personal choices and experiences but also policy, politics, and shared futures. Society, technology, scientific knowledge, everyday experience. All of these things are intertwined and interconnected, and this book is about those interconnections.

The following chapters offer a survey of relationships between science, technology, and society – not a comprehensive account (which would be impossible), but a guide to the key ways in which these relationships are articulated and how they have been described and discussed. The aim has been to bring together research that has analysed and discussed the relationship between science, technology, and society from the discipline I work within, Science and Technology Studies (STS). Over the last decades STS has investigated the intersections between expertise, democracy, activism, politics, laboratory research, technology, values, and more, offering insights that are not just helpful tools to think with, but that can help us negotiate the politics of science in society. My starting point in writing this book has been the question: what do we know about relations, intersections, and interactions between science (broadly understood) and society (also broadly understood)? I have thus tried to capture STS knowledge regarding the key dynamics, sites, languages, and practices through which technoscience and society come to interact and shape each other, from science-related activism to science in the media.

Why write such a book? Perhaps this is already clear from the way that I have described the centrality of technoscience to contemporary societies. If science and technology are ubiquitous, impacting our lives in multiple ways, it seems important to reflect on those impacts, and on the relationships between technoscience and society more broadly. Each day we are faced with questions regarding how we should handle and negotiate expert advice, emerging technologies, and scientific knowledge, from how we should weigh up competing expert accounts to whether all technoscientific progress is necessarily good, the extent to which we should understand technologies as neutral and objective, or what our role as citizens should be in engaging with technoscience. It is surely useful, as we go about negotiating such questions, to have access to research that has explored technoscience in society, and that has charted the dynamics through which science and society relate. In synthesising scholarship on these topics this book is therefore an attempt to support you, the reader, in considering the role and place of technoscience in your life, and how you might think about and engage with it.

I wrote earlier that it felt like a time of crisis.* Indeed, questions around the relationship between technoscience and society are often framed as being particularly urgent in the wake of the COVID-19 pandemic and during a renewed rise in authoritarian government. These dynamics are partly captured through what has been termed 'medical' or 'science-related populism', developments that Niels Mede and Mike Schäfer define as 'an antagonism between an (allegedly) virtuous ordinary people and an (allegedly) unvirtuous academic elite'.[6] Like other forms of populism, science-related or medical populism involves the idea that a morally superior population of 'ordinary people' are being oppressed or dominated by an elite; different, however, is that this takes place not just with regard to politics but to technoscience or medicine. The elite that is rejected or challenged is an academic one, and what is at stake is who has the right to say what 'true knowledge' is, and how this should be acted on. Examples of such populist movements include denial of the connection between HIV and AIDS in South Africa, the rejection of restrictions relating to COVID-19 in the United States, or vaccine hesitancy in the European Union.[7] More generally, science-related populism relates to a seeming disregard of facts or scientific consensus on the part of politicians such as Donald Trump in the United States or Jair Bolsonaro in Brazil, a turn that has been framed as the rise of 'post-truth' or 'post-fact' societies.

These developments are certainly shaping the lives and perspectives of many around the world. But even while distrust or questioning of technoscientific expertise may be becoming more visible or prominent, we should not

* Generally speaking, when I refer to 'crises' this is as a reflection of the ways in which particular sites or moments become labelled as such in public, such as a 'crisis of expertise' (see Chapter 7). I also use the term more colloquially, pointing to my own experiences of disruption. An exception is the climate crisis, that 'great tragedy of our time' (as Alice Bell terms it).[a] Here I am not (only) mirroring standard language about climate change in public media, but choosing to frame this as a crisis. If it is not, then what is? (See Chapter 6.)

disregard the degree to which technoscientific knowledge and expertise continue to structure how modern societies operate, nor that to which such knowledge remains in large part trusted by publics. Indeed, surveys indicate that the COVID-19 pandemic increased trust in science,[8] while trust in researchers is consistently reported as being higher than in those in most other public roles (such as journalists or politicians).[9] Rejection of expertise (whether on populist grounds or otherwise) is one aspect of the survey of the science–society landscape that this book presents, but we will also see many cases in which scientific authority continues to be a 'gold standard' for epistemic (meaning knowledge-producing) practice. Indeed, one central theme will be the diversity of the ways in which technoscience becomes visible in society. It would be impossible to find one central defining dynamic: this is also why this book is a survey, not an answer to one or more 'problems' of the relationship between technoscience and society.

A further important message before you start reading is that this book could be different. Certainly, it could be better: I am keenly aware that it is flawed in many ways. But it could also be *different*. I have written that I wanted to draw together research on science, technology, and society, and to offer this to readers as a kind of toolkit for life in a technoscientific world. One thing that has become particularly clear as I have been writing is that there is no single way of telling the story of this scholarship. I have chosen to write about particular ideas, empirical research, and themes, but someone else (or myself in a different moment or mood) may have foregrounded other things or framed the entire project in different terms, and that would have been equally valid. Indeed, this is perhaps the first – and one of the most important – lessons of the book: there is never a single right or final way of telling stories about the world. Just as histories are written, if not always by the winner, at least by people with particular perspectives on events and access to particular materials about them, scholarship is always based on particular choices. It could be otherwise.* And this is as true of scholarship in the natural sciences as it is of social research and this book in particular. What you are reading, then, is one partial and situated account of knowledge about science, technology, and society relations. It gathers together research and empirical cases that I know about and am familiar with, but my view is not total or complete.

* The notion that 'things could be otherwise' has become an STS aphorism, and is an idea that will recur throughout this book as well as in the field as a whole. One of its earliest formulations appears in Wiebe Bijker and John's Law introduction to the book *Shaping Technology/Building Society*, when they write about the contingency of technological development: 'Technologies, we are saying, are shaped. They are shaped by a range of heterogeneous factors. And, it also follows, they might have been otherwise.'[b]

Here are just a few of the choices I have made that could have been done differently: I have chosen to focus on specific sites or moments in which the social and the technoscientific become entangled, exploring spaces such as technology development, media representations, activism, disasters, expert advice, and policy and government, rather than areas of content such as health, climate science, or artificial intelligence. This means that if you are interested in any particular area (such as the climate crisis) then you will need to use the index to find the places in the book in which these are discussed so that you can navigate around it based on your interests. As I explain further in what follows, in aiming for the book to be a resource – a kind of catalogue – I expect and welcome such navigation. As a further choice, I have decided not to spend much time discussing science communication – activities that explicitly seek to share technoscientific knowledge with non-scientists, whether that is through museums, popular science writing, or festivals. While analysis of these formats is an important aspect of STS scholarship (and while I do discuss some of this work, in particular in Chapter 4), my colleagues and I have written about this in other places, and I didn't want to replicate this.[10] Finally, I have also chosen to focus on empirical cases and dynamics rather than on theoretical analysis. Again, this is certainly something that could be different: STS has a rich conceptual language that offers precision and nuance in analysing empirical contexts. I do use some of this – the term technoscience is itself one example – but for the most part this book should not be read as an account of STS theory.

The next section of this chapter goes into more detail about how (I think) you should use and understand this book. It is worth noting, then, that this first chapter is also somewhat non-traditional, in that it doesn't provide a helpful summary of the chapters that follow and of the key arguments that are made in the book as a whole. If you are looking for that (welcome, fellow skimmers!) you will find a discussion of key themes that cut across different sites in which technoscience and society interact at the beginning of Chapter 9.

Box 1.1: Central ideas and terms

While I won't provide an overview of the book's content as a whole, it does seem important to introduce some of the central terms I will be using and the meanings I am choosing to attribute to them. These key concepts are discussed at more length as they come up throughout the book, so appear here for orientation.

- *Technoscience*: I use 'science and technology' or the portmanteau 'technoscience' (which emphasises the continuities between scientific and technological practice) to refer to the methods, cultures, practices, institutions, and norms involved in (western) scientific knowledge production and technological development.* See Chapters 2 and 3.

- *(Non-)scientists*: While, as I will discuss, it is not always clear who 'belongs' and works in technoscience and who does not, I use the terms scientists or researchers to refer to those working in institutionalised scientific settings (such as universities). Non-scientists are people without such an affiliation. Depending on the context, I also use the terms citizens, publics, and audiences to refer to such people, using the plural form to emphasise that there is rarely one homogeneous public or audience. See Chapters 2 and 5.
- *Society*: Even though in practice technoscience is always entangled with the social, and distinctions between science and society are not clear, I refer to 'society' or the social when I want to specifically point to social processes, values, norms, or interactions – all of the ways in which humans relate to one another, create meaning, and make sense of the world around them. See Chapter 2.
- *Epistemic cultures and practices*: The term epistemic refers to knowledge production. Epistemic cultures are therefore communities (and their associated norms) that are oriented to making particular forms of knowledge, while epistemic practices are the routines and behaviours they use to do so.[11] See Chapters 5 and 6.
- *Epistemic diversity and injustice*: Epistemic diversity refers to the fact that multiple epistemic cultures exist – that there is a range of different approaches to creating knowledge, both inside technoscience and beyond it. The book discusses some of the ways in which technoscientific knowledge encounters knowledge from other knowledge systems, such as indigenous or experience-based knowledges. The term epistemic injustice points to the politics of such encounters, and to the ways in which particular forms of knowledge may be unfairly suppressed or silenced. See Chapter 5.
- *Expertise*: I generally use the notion of expertise to call attention to the ways in which particular people or forms of knowledge get labelled as expert in public discourse, and the ways in which this is contested, rather than as a stable or clear-cut category. See Chapter 7.
- *Politics*: In the book, politics refers not just to government and policy, but to broader debates about what collective life should look like. In the context of technoscience and society, such questions often relate to whose knowledge should count, who should make decisions on particular questions, or to the ways in which technoscience intersects with how power is concentrated and negotiated in particular contexts. See Chapter 8.
- *Justice and equity*: These terms point to questions of fairness, and to equitable access to the opportunities and benefits of contemporary societies. See Chapter 9.

---

* The term technoscience has a long history and is associated with multiple strands of STS – perhaps most significantly feminist technoscience studies (such as the work of Donna Haraway) and scholarship in actor network theory.[c]

How to use this book

It is likely that you don't need instructions for reading this book: you've made it this far, after all. Better put, there are a few more things I would like you to know about the text that follows. Feel free to skip or skim.

The first is that I have been thinking about this book as a resource (the clue's in the title: it's designed to be resources for life in a technoscientific world). In writing it I wanted to draw together key work in STS that helps reflect on the role and place of technoscience in society. This is … a lot. As a discipline, STS has been centrally concerned not only with how science is carried out and the ways it produces knowledge, but how that knowledge has moved through society. This means that any of the areas that I describe in this book, from the role of metaphors in describing science and society (Chapter 2) to the nature of ignorance and non-knowledge (Chapter 6), could fill a book in themselves (and in many cases have done so). The book therefore offers a survey of key areas of research, summarising central arguments and findings in each of these and offering cases and examples relating to them, but rarely going into a lot of detail. Instead I have included references to key work that I discuss, but also to texts that offer extra detail, additional perspectives, or contrasting views. I therefore anticipate that readers will indeed use the book as a resource, a starting point from which you can travel to other literature and engage with more in-depth discussion. My hope is that the book offers not just resources in the sense of key ideas with which we can think about the relationship between science and society, but also in being a means of identifying authors and work with which to explore these ideas further.

This relates to another aspect I find it important to flag. I have mentioned my home discipline, STS, and the fact that I am trying to synthesise research from it on technoscience and society. However, to me what characterises STS is not just what it studies – the processes and practices through which technoscience is made[12] – but the ways in which it does so. In researching scholarly knowledge production it is also studying itself, and the field has therefore come to incorporate the notion of reflexivity, the importance of reflecting on and analysing one's own research practices. While I wanted, in writing this book, to present an account of STS knowledge on technoscience and society, I therefore found that I constantly needed to reflect on and situate the status of this knowledge. This is why I have already spent quite a lot of time talking about myself and my choices, and why I will continue to incorporate a degree of reflection and commentary on the text. If a central theme will be that, to quote the subtitle of a book by Steven Shapin, technoscience is 'Produced by People with Bodies, Situated in Time, Space, Culture, and Society, and Struggling for Credibility and Authority',[13] it seems essential to acknowledge that I am in exactly the same position: even if I at times adopt a rather abstract, authoritative authorial voice, presenting the ideas I discuss as if they emerged from nowhere and attempting to summarise and synthesise them, my body, experiences, and situation are shaping what I write.

I therefore do my best to highlight the ways in which (my) knowledge and arguments are situated and contingent. I often name the people whose work I am discussing, for example, to emphasise that it has indeed emerged from particular individuals and places. One important thing to bear in mind is that I am a White,* cis-heterosexual woman who comes from the United Kingdom, and has worked at universities in the United States, Denmark, Norway, and now Austria. While I read some Danish and German, my professional and personal experiences have been oriented to Europe and to Anglophone contexts. In conjunction with the fact that many 'classics' of STS analysis have tended to be similarly Euro-American in the cases and contexts they engage with, I have found myself frustrated and embarrassed by the limited range of examples or sites I find it easy to draw upon.† While I have sought to extend these as much as possible – in particular with the help of my colleagues Nora Ederer and Elaine Goldberg, who have helped me research a broader range of cases than is found in the scholarly literature – this book can only be a partial account, one shaped (again) by my situatedness. What I want to avoid above all else is the idea that this partial account is a universal one, reproducing a particular perspective as somehow comprehensive. Please do read with this in mind.

This emphasis on situated knowledge has also shaped the format of the book that you are reading. You will have already noticed that I use different slightly kinds of text within it. The main text (which is what you are reading now) carries the central discussion; I write in the first person (obviously), but this material offers an overview of central dynamics and themes so is often rather general, describing key findings and arguments according to the scholarly literature. I provide references to that literature in endnotes, which can be found at the end of each chapter; again, I hope you are able to use these to read my account in dialogue with the wider discussions I am pointing to. In addition I use footnotes in places where I find it important to provide some additional commentary – for instance when I want to situate particular knowledge claims, reflect on my own position, or provide some additional information that is not relevant to the main flow of the text.‡ References within footnotes are marked with letters, and appear in a section before the main endnotes at the end of each chapter. Finally, text boxes appear throughout. These generally include

---

* In line with contemporary guidance, I capitalise White and Black to signal that these are not neutral descriptors but historically created identities.

† Indeed, analysis of STS as a field shows that it is a 'white space' both in terms of a continuing lack of diversity and with regard to its minimal engagement with race, racism, and Black scholarship.[d]

‡ In using footnotes in this way I have been inspired by the writing of authors such as Annemarie Mol and Max Liboiron – both of whom use different kinds of text to situate

concrete examples or cases of the issues or dynamics described in the main text, but they sometimes clarify terminology or offer additional reflection, and it is these that have been the focus of my efforts to illustrate central ideas through a more diverse range of empirical sites and cases.*

It is now a cliché to say that a book is never finished, only abandoned. In sending you on your way into the next chapter I do, however, want to say that I view this book not as something complete, but as a starting point: my plan is to use it as a structure for collecting further cases, examples, and literature on the myriad ways in which technoscience is instantiated across diverse societal settings. I hope that it is similarly a starting point for further reflections for you.

Conclusion

This book offers an introduction to and overview of the ways in which science, technology, and society are entangled, as discussed in Science and Technology Studies. Each chapter explores particular spaces or processes in which these entanglements take shape, describing what is known about these as well as cases and examples of them. The next chapter is, however, a little different in that it continues setting the scene, asking: how do we talk about 'science and society', and what are the histories of relationships between them?

References

[a] Bell, A. (2021). *Our Biggest Experiment: An Epic History of the Climate Crisis*. Catapult, p 18.

[b] Bijker, W.E. and Law, J. (1994). *Shaping Technology/Building Society*. MIT Press, p 3.

[c] McNeil, M. (2008). *Feminist Cultural Studies of Science and Technology*. Routledge. Latour, B. (1987). *Science in Action: How to Follow Scientists and Engineers through Society*. Harvard University Press.

[d] Mascarenhas, M. (2018). White Space and Dark Matter: Prying Open the Black Box of STS. *Science, Technology, & Human Values*, 43(2), 151–170.

[e] For the former, see: Liboiron, M. (2021). *Pollution Is Colonialism*. Duke University Press. Mol, A. (2002). *The Body Multiple: Ontology in Medical Practice*. Duke University Press.

---

and reflect on their arguments – but also the comic fantasy writer Terry Pratchett, who was the first to show me the sly power of a footnote when I read him as a teenager.[e]

* I am not fully satisfied with this as a solution to a non-diverse literature, because often what happens is that I describe key dynamics or concepts that have emerged from Global North scholarship in the main text, and illustrate them with cases or examples from elsewhere. I should be clear, then: there is plenty of scholarship coming from the South, but my limited perspective means that this is less accessible to me, and I don't necessarily have the knowledge or right to explain it to others.

[1] Chivvis, C. (2022, 3 March). *How Does This End?* Carnegie Endowment for International Peace. https://carnegieendowment.org/2022/03/03/how-does-this-end-pub-86570

[2] Service, R. (2020, 29 September). One Number Could Help Reveal How Infectious a COVID-19 Patient Is. Should Test Results Include It? *Science*. https://www.science.org/content/article/one-number-could-help-reveal-how-infectious-covid-19-patient-should-test-results

[3] Harvey, H. (2022, 9 March). Carbon Dioxide Will Have to Be Removed from Air to Achieve 1.5C, Says Report. *The Guardian*. https://www.theguardian.com/environment/2022/mar/09/carbon-dioxide-removed-from-air-carbon-offset-market-report

[4] Taylor, M. (2022, 7 March). Six Promises You Can Make to Help Reduce Carbon Emissions. *The Guardian*. https://www.theguardian.com/environment/2022/mar/07/six-promises-you-can-make-to-help-reduce-carbon-emissions

[5] IPCC (2022). *Climate Change 2022: Impacts, Adaptation and Vulnerability: Contribution of Working Group II to the Sixth Assessment Report of the Intergovernmental Panel on Climate Change*. https://www.ipcc.ch/report/ar6/wg2/

[6] Lasco, G. and Curato, N. (2019). Medical Populism. *Social Science & Medicine*, 221, 1–8. Mede, N.G. and Schäfer, M.S. (2020). Science-related Populism: Conceptualizing Populist Demands toward Science. *Public Understanding of Science*, 29(5), 473–491, at p 484.

[7] Mede and Schäfer (2020). Stoeckel, F., Carter, C., Lyons, B.A., and Reifler, J. (2022). The Politics of Vaccine Hesitancy in Europe. *European Journal of Public Health*, 32(4), 636–642.

[8] Wellcome (2021). *Wellcome Global Monitor 2020: Covid-19*. https://wellcome.org/reports/wellcome-global-monitor-covid-19/2020

[9] *Nature* (2024). How Can Scientists Make the Most of the Public's Trust in Them? *Nature*, 626(7997), 8–8. https://doi.org/10.1038/d41586-024-00238-x

[10] See Davies, S. and Horst, M. (2016). *Science Communication: Culture, Identity and Citizenship*. Palgrave Macmillan. Felt, U. and Davies, S.R. (eds) (2020). *Exploring Science Communication*. SAGE.

[11] Knorr-Cetina, K. (1999). *Epistemic Cultures: How the Sciences Make Knowledge*. Harvard University Press.

[12] Felt, U., Fouché, R., Miller, C.A., and Smith-Doerr, L. (2017). Introduction to the Fourth Edition of the Handbook of Science and Technology Studies, in Felt, U., Fouché, R., Miller, C.A., and Smith-Doerr, L. (eds) *The Handbook of Science and Technology Studies* (4th edn). The MIT Press, pp 1–26.

[13] Shapin, S. (2010). *Never Pure: Historical Studies of Science as if It Was Produced by People with Bodies, Situated in Time, Space, Culture, and Society, and Struggling for Credibility and Authority*. Johns Hopkins University Press.

2

Histories and Imaginations

R.F. Kuang's 2022 novel *Babel* is a fantasy, set in a world in which the act of translation between languages can spark magic. But it is also a meticulously researched account of Oxford in the 1830s, a world in which knowledge production is entangled with the maintenance of colonial power. At the novel's centre is the Royal Institute of Translation, housed in a tower in the centre of Oxford – known as 'Babel' – that is eight stories high and the tallest structure in the city. The home of the translation work that powers the British Empire, it is a magically protected 'gleaming white edifice' accessible only to an elite cohort of scholars. It is the heart of translation scholarship, and thereby – in the logic of the novel – the seat of power that is used to exploit and dominate the world.

Intentionally or not, Kuang's Babel echoes images of academia as an ivory tower, a metaphor used to present universities as cloistered environments that are segregated from the societies in which they sit. As Steven Shapin has charted, the use of this metaphor intensified throughout the 20th century, increasingly becoming attached to research and researchers and being used to critique an attitude of academic detachment.[1] 'This is no time for any man [*sic*] to withdraw into some ivory tower', Shapin quotes US President Roosevelt as saying in 1940, 'and proclaim the right to hold himself aloof from the problems, yes, and the agonies of his society'. The image of the ivory tower continues to circulate and (Shapin suggests) is now almost entirely framed as negative, capturing the idea that research that is disengaged from society and its needs is morally problematic. In both *Babel* and wider discourse, ivory towers are dangerous.

To think of the relationship between science and society is, often, to reach for metaphors such as the ivory tower. Such languages are taken-for-granted means of expressing how academic knowledge production relates to wider society, and the assumptions that lie behind them are perhaps too rarely explored. In this chapter I would like to do so, and to situate and explore the things we call 'science' and 'society'. In multiple ways the discussion here sets the scene for the chapters that follow, in particular by introducing the central thesis of the book: that science, technology, and society are not pre-existing spaces or activities that can be clearly distinguished, but are always intertwined. Ivory towers are never as cloistered as they might appear. In the examples that follow we will therefore examine some of the ways in which these entwinements and interactions have been constituted, throughout history and in the contemporary world.

Science has always been public

One way of exploring the entanglement of science and society is to look at the history of western science, and the ways in which this has developed through the involvement of different public audiences and participants.* Take, for instance, the problem facing the earliest practitioners of what became understood as 'science' in 17th-century Europe. How could the knowledge that they were producing using their (rather new) methods and approaches be considered robust and reliable? Or, to put it another way, why should anyone believe what they were saying about their findings? This was especially important in a context in which the forerunners of these early scientists had been alchemists, magicians, and theologians, most of whom wrote in languages other than those used by the majority of the population (such as Latin) and some of whom exactly framed their authority as based on their access to hidden, arcane knowledges. In such cases their secrecy, their separateness from wider society, was their selling point. This approach was antithetical to the ideas of 17th-century empiricists such as Francis Bacon or the founders of Britain's Royal Society. These scholars, James Hannam writes, 'saw science as analogous to law. When they did experiments, they wanted them to be witnessed as a way of validating what happened'.[2] Public witnesses were therefore essential to 'proper' scientific practice; later, 'literary technologies', or forms of writing, were developed that closely described, and allowed a reader to replicate, those experiments.[3] Witnessing, whether virtual (through written accounts) or in person (when one visited a demonstration or experiment), was thus central to how modern science was conceived – though it is important to note that in practice those witnesses were a 'carefully selected and disciplined public'.[4] Not just anyone could vouch for the trustworthiness of an experiment: to be authoritative, one should be White, educated, male, and a 'gentleman'.

* This is, of course, just one history of science – one that is frequently told, and that is useful for my purposes here (exploring the relation between 'science' and 'society'), but that should be understood as partial given the long histories of knowledge production in non-western contexts. An 'origin story' of what we call science today could just as well begin with astronomy in China, Māori navigation, or Ancient Egyptian mathematics.[a]

Box 2.1: Histories and counter-histories of science

The history I describe here is one that follows the development of a set of techniques by (mostly) wealthy 17th-century men, and that frames these techniques as circulating to other contexts. This is, of course, just one history that could be told both of the development of western science and of knowledge production more generally. Work in the 'global history' of science has emphasised that scientific practices emerged around the world, and have been circulating and intermingling for millennia. Even European science was 'a diverse and heterogeneous body of knowledge, institutions, and practices' which owed much to 'multiple, including non-Western, traditions of knowledge'.[5] Others have argued for the need to tell 'counter-histories' of science that show how the achievements of western scholarship were often grounded in the knowledge and resources of poor workers or colonised peoples.[6]

Science has also always been public in the sense of relying on support from non-scientists. While many early scientists were independently wealthy, and much science was for many years carried out by amateurs and hobbyists,* others, like mathematician and astronomer Galileo Galilei, had to fight for patronage from wealthy benefactors. (Hannam describes how Galileo's efforts at public communication of his work had the desired effect of rendering him a local celebrity, helping him get a well-paid job with the Duke of Tuscany that allowed plenty of time for research.)[7] If one challenge for scientists has been the question 'Why is your knowledge authoritative?', another is therefore 'Who will pay for it?'. As well as finding wealthy patrons, one approach has been to become – in essence – a public entertainer. There is a long history of public demonstrations, shows, and educational books which serve the double role of explaining (and justifying the value of) scientific work and raising funds to further support it. Joseph Wright's dramatic 1768 painting, 'An Experiment on a Bird in the Air Pump', shows one such demonstration in a wealthy private home;[8] in it, a travelling lecturer uses the newly developed air pump to remove the air from a flask with a bird in, while the extended family looks on with reactions from horror to fascination or indifference. Such performances also took place in public venues across society. Science was a spectacle, to be enjoyed by both working-class audiences (who might

* There is a long tradition of amateur practitioners of science, for instance in the context of natural history. Science carried out in people's free time was considered just as valid as that done by professional scientists.[b]

be entertained by displays of electricity or mesmerism) and more elite groups (who could visit a theatre or opera house to watch astronomy lectures and displays)[9] – and producing such spectacles helped fund the research that they were based on.

Early European scientists thus relied on public witnesses both for the legitimacy of their scholarship and for its funding and support. In addition, non-scientists – individuals or groups who would not identify themselves as doing or working in science – have often been essential to *making* scientific knowledge. One example is the role of technicians in scientific research. Today, technical staff work in laboratories (or other spaces where research is carried out) to enable and assist research: they might order and organise essential materials, care for equipment, build or maintain machines, tend to animals, prepare samples, or carry out experimental procedures. As this list suggests, this is skilled work, often involving years of on-the-job (or formal) training and experience. Such labour is not a recent phenomenon: science has always relied on the activities of those who clean, assist, mend, take notes, or provide materials. Steven Shapin, for instance, paints a picture of the laboratory of early scientist Robert Boyle as a bustling space through which servants, assistants, and tradespeople moved according to the dictates of 'the Master'.[10] These early scientists were used to being waited on, and took for granted the assistance of domestic and other servants. Historians of 20th-century science trace a direct line between such 17th-century servants and the contemporary role of the technician: 'the modern laboratory technician', writes Tilli Tansey, 'has evolved from the tradition of personal laboratory servants, assistants and amanuenses employed in the laboratories of seventeenth century natural scientists'.[11]

One unifying feature of technicians past and present is that they are largely invisible in written records of science. Shapin notes that early scientific assistants only become visible in scientific archives when something went wrong, at which point they were blamed for their faulty experimental practice (ironic, given that many scientists rarely carried out the work of experimentation themselves, instead directing others). Even in the 20th century, technicians and their work 'remain difficult to see':[12] they rarely appear as authors on research papers, and their activities generally remain behind the scenes, even though their experience and expertise may be central to getting experiments to work.[13] One central counter-example is the creation of Dolly the sheep, the world's first mammal cloned from an adult cell, who was born in 1996 in Scotland. An employment tribunal a decade later explored both questions of scientific credit – which scientist was most responsible for the result – and the lack of acknowledgement of the work of the technicians involved, who had carried out the 'intricate and arduous egg and cell manipulation needed to create each clone'.[14] Technicians Bill Ritchie and Karen Mycock argued that they should have appeared as authors on the high profile research article published about the experiment. While the case sparked debate about what kind of work 'counts' within scientific research, and how different forms of labour can be credited,[15] technicians often continue to be invisible within accounts of science. Indeed, thinking about how to acknowledge and credit the work of skilled technicians and assistants opens up broader questions about the other people involved in supporting, and providing an environment for, research. Science is enabled not just by technicians but by administrative personnel, students, communicators, cleaners, porters, and a myriad of others who are active in universities, businesses, and research organisations. The work of all of these non-scientists is essential to scientific practice.

If the example of technicians shows us how the work of non-scientists is present within, and enables, research, a further example highlights the exploitative form that this may take. It is clear that contemporary science has emerged out of a set of practices that not only involved non-scientists, but that violently extracted value from them, and that these histories continue to shape technoscience today. Historians and postcolonial scholars have shown how modern science was advanced by, but also intrinsically tied to, colonialism. Research was funded to aid colonisation and to assist in the management of the colonies: technologies such as the telegraph became important as a means of reaching, and governing, colonial holdings, while research fields such as botany, ecology, or tropical medicine became salient (and fundable) as new lands needed to be understood and exploited.[16] Key advances in scientific knowledge were only possible because of scientists' access to the materials and knowledges of colonised lands. As Suman Seth writes, '[o]ne cannot imagine Charles Darwin's work being possible without his access to plant and animal specimens derived from several European empires'.[17] Similarly, racist assumptions were baked into the work of scientists, whether that meant categorising the peoples encountered during colonial exploration or implicitly or explicitly participating in the slave trade. 'Wherever these colonialist scientific expeditions went', write Subhadra Das and Miranda Lowe, reflecting on the history of the natural history collections now celebrated in museums, 'subjugation of native people, slavery, and genocide were the result'.[18] As such expeditions expanded, researchers made use of the knowledge of the peoples they encountered without credit, or used them as assistants in a similar way to the technicians described earlier.

19th- and 20th-century science (in particular) was thus not only driven by the concerns of colonising societies, but fundamentally relied on extracting resources from colonised sites. Publics around the globe became more or less unwilling participants in processes of sample collection or analysis, from themselves providing the material for 'race science' to the unacknowledged use of indigenous knowledges in western science and technology.

Box 2.2: Re-examining colonial histories of invention

Histories of science and technology have often been triumphal in nature, focusing on the inventions or discoveries of key individuals. Increasingly, however, historical research is indicating the more complex stories behind these accounts, and the ways in which discovery or innovation is collective, shaped through use as much as by specific moments of invention. Jenny Bulstrode explores one celebrated 18th-century invention, a 'process of rendering scrap metal into valuable bar iron that has been celebrated as one of the most important innovations in the making of the modern world' that was patented by 'British financier turned ironmaster' Henry Cort. In examining the background to the patent application, Bulstrode shows how this invention (framed as central to the industrial revolution) was based on a long history of skilled metal work in West and West-Central Africa, the innovative work of Black metallurgists based at an iron foundry in Jamaica run by an enslaver, and Cort's appropriation of their work. 'Cort had "found out" how to transform bundles of scrap into valuable bar iron', she writes. 'But John Reeder's now dismantled works had achieved this mechanical alchemy years before, with scrap iron, reverberatory furnace and rolling mills, and the skill of the 76 Black metallurgists.' Bulstrode's aim is not (only) to document theft of intellectual work, but to examine the ways in which this invention was grounded in the Black metallurgists' traditions and meanings.[19] In the 'smithing lineages' the enslaved iron workers were rooted in, 'iron had often embodied both principles of the sacred and of the profane. Against the onslaught of Europe's human trade, working iron was a means of expression to forge fighting alliances, heal sickness and express grief'.[20]

Technoscience and society in the contemporary world

We see the echoes, effects, or reproduction of many of these dynamics from the history of science today, such that science continues to be a public activity, and society continues to shape science. Extractive practices such as those intrinsic to colonialism continue in various forms. For instance, bioprospecting – trawling through regions and organisms in the hope of finding biologically active substances that might be used in product development – can exploit the knowledge, heritage, or lands of indigenous communities.[21] Large-scale studies of genomes or other aspects of populations may rely on data or biological material that individuals have no idea has been collected or will be used in research.* And the rise of big data scholarship in everything from communication studies to the development of artificial

* The 'CARE principles' have been one response to this.[c] See discussion in Chapter 8.

intelligence relies both on huge amounts of material collected from publics (via their social media presence, for instance) and on the low-paid, rarely acknowledged activities of cloud workers.[22] Here, once again, science and technology are reliant on the activities, bodies, or practices of publics in ways that extract value from those publics, while simultaneously rendering them invisible. The involvement of non-scientists in scientific work, the use of public communication for promotion and fundraising, and an emphasis on openness and transparency are similarly all familiar features of contemporary science.

The novelty of emphasising these interconnections between science, technology, publics, and wider society shouldn't be exaggerated. Technoscience and society are, after all, intertwined in many obvious ways that we are happy to take for granted. It is clear that science and technology respond to human needs and directives – during the COVID-19 pandemic in the early 2020s, for example, research and researchers rapidly oriented themselves to understanding and intervening in the virus and its progression, and as a result vaccines were developed in record times.[23] We also take for granted that the ways in which we live our lives and think about the world are impacted by scientific knowledge and the technologies that we use (an idea we will return to in Chapter 3). Advances in technoscience allow us to fly across the world in hours, communicate with those far away as though they were present, and to approach childbirth or infectious disease with much less fear than would have been imaginable some centuries ago. Such possibilities certainly impact our thinking, behaviours, and day-to-day lives.*

The rest of the book can be understood as an extended exploration of this mutual shaping of science, technology, and society, in that it discusses different sites and contexts and looks at how technoscience and society are made together within them. However, it is worth already exploring some of the more subtle ways in which human practices shape technoscientific knowledge, and vice versa. One example is of the ways in which cultural norms and assumptions may come to shape the content of science, even that which is represented as objective or value-free. For instance, Emily Martin has analysed the language used to describe reproductive processes – and in particular the interactions between egg and sperm cells – in scientific texts.[24] What she finds is a startling reliance on gender stereotypes to describe the different kinds of cells: 'It is remarkable', she writes, 'how "femininely"

* Though, importantly, these benefits of technoscience are not equally distributed. Such advances have changed how we think about and encounter the world, but cultural structures and norms continue to shape how they are accessed and utilised – which is exactly an example of how technoscience is never encountered in ways separate from wider society.

the egg behaves and how "masculinely" the sperm'.[25] The use of such stereotyped language (eggs are passive, dormant, and fragile; sperm active and vigorous in 'penetrating' an egg) has shaped not just how reproduction is discussed, but how research into it is carried out. Martin argues that the 'sleeping metaphors' of gender stereotyping have actively mitigated against science that, for instance, reveals the egg to be a more active agent within interactions with sperm than had previously been thought. Even in what purports to be the most neutral descriptions of biological processes, wider cultural values play a role in what can be seen, described, and become a fact.*

Other scholarship has discussed how, for example, a wide range of choices come into play when reporting causes of death or prevalence of particular diseases: it is not always clear what 'counts' as a death, how to classify causes of it, or what the correct diagnosis of a medical condition is. Individual judgements – which may be influenced by cultural norms and values – about all of these rapidly become integrated into statistics concerning death and morbidity, which are thus shaped by a set of contingent human choices.[26] Even the so-called 'hard case' for arguments of the social shaping of knowledge, research in pure mathematics, can be understood as constituted by the societal context of the research, the language choices of its participants, and hierarchies and norms within the discipline.[27] Value judgements, priorities, and assumptions co-constitute technoscience, in ways that may remain invisible until they are contested or protested.†

Box 2.3: Gender and racial bias in technoscience

Are medicine and medical technologies neutral? Scholarship has shown that an imagined user is central to how medical drugs and devices are developed, and that this user often continues to be framed as a White man. For example, the pulse oximeter is a device that helps to measure blood oxygen levels which became hugely important during the COVID-19 pandemic. It had largely been tested on light-skinned people; for non-White people, it is much less accurate.[28] Similarly, it is clear that a 'data gap' exists around women, with many aspects of women's lives not having been framed as important or interesting enough to research.[29] In such examples, White, male bodies and experiences are framed as universal. The result is that anything from piano keyboards to drugs are

* This is just one example of a much larger body of research that has examined how science is shaped by social practices and values. We'll encounter more of this throughout the book.[d]

† This should not be misunderstood as meaning that it is *only* social processes and factors that constitute technoscience. Rather we can understand knowledge production and the development of technologies as collaborations between the material world, scientific apparatuses and systems, individuals and collectives, wider social contexts, and much more.[e]

often designed for a (particular kind of) man's body, causing discomfort and at times danger to anyone outside of this model.[30]

So social processes, cultural norms, and value judgements all play a role in constituting scientific knowledge production. But science and technology also have very real effects on societies and on individual lives, in ways that go beyond the obvious. Take, for example, map-making, the process of representing – and imposing names and order upon – geographical landscapes. Mapping technologies formed part of a suite of technosciences developed in order to assist with European colonisation of the Americas, Africa, and Asia. Maps were used not just to describe geographical features, but to lay claim to the land, carve it up between different groups, and give new, settler-imposed, names to the landscape. Cartography involves state-of-the-art knowledge (for instance with regard to features such as height or depth) while simultaneously being a deeply political practice. In the context of settler colonialism, map-making has historically removed the languages, meanings, and land relationships of indigenous peoples from the record, while imposing at times arbitrary borders (consider the straight lines that divide the states in the United States) that go on to shape opportunities and possibilities for those on one side or the other. Technical decisions about precisely where borders lie or how terrain should be described thus have profound effects on how inhabitants of a particular region name and make sense of it, or on citizenship and rights.[31] Similarly, recent research on ocean mapping shows how such processes highlight some aspects of the oceans over others, foregrounding particular kinds of data (shipping routes rather than weather events, for example) and thereby shaping how marine environments are understood and engaged with by policy makers.[32] Aspects of technoscience such as maps – or other knowledge or technological products – have impacts on the world, remaking it in their image or shaping the possibilities available to individuals or groups.[33]

Box 2.4: Technologies of time

Many aspects of technoscience are so taken for granted that the ways in which they embed particular assumptions or encourage us to act in particular ways may be invisible to us. For instance, Ingrid Erickson and Judy Wajcman write about how digital calendars are developed, noting that such tools are 'so integrated into our lives via mobile phones and other devices that we rarely think about how they may be shaping the tempo and pace of everyday life'.[34] Indeed, they find that digital calendars are created by Silicon Valley software designers with productivity in mind, and that they focus on 'auditing' the 'asset of time' in order to assist users in self-optimisation. While there is nothing wrong with such an approach, it is just one of many ways of thinking about how we use time. 'Silicon Valley appears to be both reifying and reinforcing a singular construction of "optimal" that is centered on efficiency and cost', write Erickson and Wajcman.[35] Importantly, '[n]on-productive forms of time use (e.g., care work, play, pottering) do not seem to fit this paradigm'.[36] Our calendars may therefore be shaping our expectations about time use, and our behaviours as we plan and use time, in ways that exclude aspects of life that may in fact be important (such as 'care work, play, pottering').[37]

Imagining the science–society relationship

Engagement with public data, audiences, and interlocutors – the mixing of the scientific and the social – thus is and always has been constitutive of scientific practice. In this book I will therefore assume that 'science' and 'society' are not separate entities, but are – in different ways and to different degrees – entangled with each other. As noted, this is in many ways obvious, even banal. At the same time, it is important to emphasise this mutual shaping given the languages and metaphors that are commonly used to discuss science and society. As well as the ivory tower, we find a range of terms relating to gaps, chasms, and bridges across these – a language that frames science and society as separate realms that must be brought together. The notion of 'bridging the gap' is frequently used to convey the urgency of crafting these connections, with the term being used to describe science communication activities from science policy events[38] to citizen science projects[39] or student placements in government organisations.[40] To quote a scientist involved in one such activity, the goal was 'to facilitate a dialogue between the sciences and society' and to 'strengthen the bonds between the world of research and the general public'.[41] This language of separation, of gaps between distinct worlds and the need to narrow them, is so common as to have become almost invisible. To Bernadette Bensaude-Vincent, however, such metaphors speak to a false dichotomy between researchers and laypeople: 'what follows [from this language] is a polarization of the social distribution of knowledge within modern societies. … Public knowledge is denied all relevance while a minority of scientists holds the monopoly of legitimate knowledge'.[42] Something similar is implied by a further set of metaphors that circulate around the science–society relationship. Here we find the language of dissemination or enlightenment: 'Science Finds, Industry Applies, Man Conforms', to use the motto of the 1933 Chicago World Fair.[43] Knowledge moves from the world of science outwards, to affect and shape society. Such language presents science as a kind of sun, which radiates knowledge (and technologies) out into the world; we benefit from this, but (as with the sun) have little say in how it happens.

Box 2.5: Traditional knowledges and publics

While many of the basic metaphors and ideas used in western science to talk about the relation between knowledge production and wider society emphasise separation and distinctiveness, this is not the case for all knowledge systems. Many indigeneous knowledges take a holistic and contextual approach to knowledge, presenting ways of knowing as always intertwined with specific processes, places, circumstances, and actors. For instance, environmental philosopher and indigenous scholar Kyle Powys Whyte writes about what is sometimes referred to as 'Traditional Environmental Knowledge': these are, he notes, 'systems of responsibilities that arise from particular cosmological beliefs about the relationships between living beings and non-living things or humans and the natural world'.[44] Knowledge cannot simply be extracted from its context of creation and use, but implies particular responsibilities and relationships. In such perspectives to talk of 'science' and 'society' as separate is even more nonsensical than in the context of western science.

There are, of course, other metaphors in use that describe particular aspects of science, from research as a journey to the many metaphors that seek to describe particular scientific phenonema (the brain as a computer, DNA as a book, discoveries as 'breakthroughs').* Those that I have just mentioned are significant, though, in that they speak to the overall relationship between science and society and thereby present implicit models of that relationship. Metaphors function as heuristics, guides for how we collectively think about particular topics or issues, and are therefore more important than we might initially imagine – Sally Wyatt, for example, discusses how the metaphors that circulate around digital technologies are 'a resource deployed by a variety of actors to shape the future'.[45] Metaphors about science and society can thus shape how we think about technoscience and its relation to society by framing those entities (and how they relate) in particular terms. While the metaphors I have discussed emphasise different features of how research and wider society relate, they also have some commonalities, the most important of which is that they present science and society as *separate*. The ivory tower suggests seclusion and detachment; bridges link different

* For those interested, Brigitte Nerlich has extensively explored the role of metaphors in science.[f]

domains; enlightenment and dissemination suggest distinct sources and receivers of knowledge. These taken-for-granted ways of imagining and talking about science and society therefore suggest that there are, indeed, separate things called 'science' and 'society'. (Of course, so does the way that I have been writing. Even to use these different terms suggests that they are distinct entities.)

As I have noted, however, the history of western science makes it clear that research has never been cut off from non-scientists. In this book I therefore write about the place of science within wider society in a way that acknowledges the entanglement of these domains. Science is always social, shaped by and part of wider economic, social, political, and institutional dynamics; at the same time, we can think about modern societies as intrinsically technoscientific, defined by their use of technologies and authoritative research-based knowledge. Science and technology shape our lives and what is possible (an idea sometimes known, in its strong form, as technological determinism – something I will discuss further in Chapter 3), while simultaneously being shaped by human actions, values, and choices. Rather than talking of science and society as separate entities, perhaps it would be better to speak of 'sociotechnoscience', if doing so were not quite so unwieldy.* Importantly, these intersections and interactions are not stable, or look the same everywhere. What is interesting is to look at – and what we will explore by engaging with different kinds of sites and practices throughout this volume – is how these entanglements take place in particular contexts.

Box 2.6: A note on language

I will not use the term 'sociotechnoscience': it is just too clumsy. Instead I talk about 'science and technology' or 'technoscience' (to emphasise the continuities between scientific and technological practice) to refer to the methods, cultures, practices, institutions, and norms involved in (western) scientific knowledge production and technological development. My view is that science is not the only source of knowledge, so I also discuss knowledge from other sites or communities, such as indigenous or experience-based knowledges. And even though in practice technoscience is always entangled with the social, I talk about 'society' or the social when I want to specifically point to social processes, values, norms, or interactions – all of the ways in which humans relate to one another, create meaning, and make sense of the world around them.

---

* Finding a language that acknowledges the co-constitution of science and society is challenging. Sheila Jasanoff uses the notion of co-production; Irwin and Michael speak of 'ethno-epistemic assemblages' in the context of specific entanglements of the social and scientific.[g]

Technoscience and the public good

The rest of this book thus explores ways and places in which 'science' and 'society' are mutually constituted. As we will see, many of these sites and interactions involve politics – questions about whose knowledge should count, or who should make decisions on particular questions. Indeed, if there is a long history of science relying upon the support and knowledge of non-scientists, there is almost as long a history of reflection on the role that it should play in the societies it is located within. In the rest of this chapter I therefore consider some of the ways in which technoscience has been discussed in the context of the public good, democratic society, and governance (themes I will pick up on further in Chapter 8, in particular).

The history I recounted earlier emphasised the need for early researchers either to be independently wealthy or to find patrons or other forms of support (for instance by carrying out public displays). Later years, however, increasingly found science framed as a 'goose that lays golden eggs' that should be funded by the state. As Steven Shapin writes, '[s]ectors of the seventeenth century English public came to believe that science might possess answers to economic and military problems they desired to solve'.[46] Research was viewed as being potentially useful to governments, states, and citizens, and accordingly as something that should be funded and organised. The 19th century, in particular, started to see the institutionalisation of public support for research, and as universities increasingly received financing from national governments the notion of an implicit 'social contract' began to form. In the Humboldtian university model that spread from Germany, university scholars were asked to carry both research and teaching, and thereby to combine their scholarly activities with the education of (good) citizens.[47] The notion of a social contract between science and society reached its peak in the postwar period of the 20th century in the United States, where science received large amounts of funding based on the promise of the technologies and social progress that its activities would ultimately result in.[48] Often, such funding was implicitly or explicitly tied to military ambitions, to the extent that Shapin wrote, in 1990, that it 'is now difficult to imagine what the social institution of science would look like divorced from its military ties'.[49] While these dynamics have shifted, and the implied linear relationship between basic research, innovation, and societal impacts has been repeatedly questioned (as discussed further in Chapter 3), an imagination of a social contract between science and society continues to circulate. In this view publicly funded science should in some way result in societal benefits – and thereby be responsive to public needs, values, and desires.[50]

This view has been expressed in different ways in different moments and contexts. The 20th century saw concerted efforts from academics themselves to try and render science more open and accessible, both in the context of political science and in the activities and concerns of natural scientists. Take, for instance, the activities of 'radical science' movements from the 1960s onwards. As Alice Bell has described,[51] those studying and working in science were not left untouched by the protests and upheavals of the late 1960s. At a time when young people around the world were calling for equal rights and an end to war, some scientists also began to frame themselves as radicals, and to reject what they saw as the conservative norms of contemporary science. The British Society for Social Responsibility in Science and the US organisation Science for the People[52] were both founded in 1969, and rapidly became active in, for instance, protesting the use of science for military purposes, supporting women in science, and helping workers to monitor and fight against dangerous conditions. Those involved in such movements were concerned about how science was funded and governed, and about the ways it was being used (in chemical and biological warfare, for instance). But they were also passionate about its potential when turned to public good, and experimented with 'community science' projects where those with scientific and technical skills would use these to help document or respond to public problems such as air pollution, or the need for medical advice. A 1970 edition of the magazine *Science for the People* calls for participants in its 'Technical Assistance Program', which had a charter to 'to assist community political groups in situations where technical experience and knowledge can make their struggle more effective', as well as helping 'demystify' technology.[53] Such activities responded to public needs, and framed science as something that should be in service to people. As Bell writes:

> Science, BSSRS [the British Society for Social Responsibility in Science] believed, was humanity's greatest hope, but it was also becoming dangerously corrupt. Science could change the world, but it also needed to change itself … the core of BSSRS were 1970s radicals schooled in consciousness-raising women's groups and anti-war sit-ins. They had a different attitude to science, the state and ideas of authority.[54]

At its height, radical science had its own journals, and staff members who supported progressive action and public reflection on science's social responsibilities. But the organisations faded out in the 1980s and 1990s: victims, Bell suggests, of the rise of individualism and bullish market capitalism.

Box 2.7: Technoscience and the military

One of the things that many radical scientists were concerned about was science and technology's entanglement with the military, and the way in which research and development were shaped by the needs and interests of armed forces. Writing in

1970, Hilary and Stephen Rose describe how the two world wars of the 20th century accelerated research into chemistry and physics, and took it in directions that would support military power.[55] Military funding has been critical to technoscience over the 20th and 21st centuries, and while there has been much debate regarding what this does to scientific research, researchers such as the Roses emphasise the need to recognise this entanglement and to reflect on what responsibility means in such contexts.[56] As Jane Gregory writes:

> The Cold War was the context and engine for much of the science and science communication of the [20th century], until capitalism (war by other means) took over as our dominant driver. ... Like it or not, we [scientists and science communicators] play an active part in this global system, and we should be alert to it.[57]

The aims of the radical science movement – to ensure that scientific research was used to support the public good – were also present in more institutionalised experiments and structures. One example is the Science Shop Movement, which, similar to the community science initiatives described earlier, aimed to bring laypeople and researchers together such that research could help answer public problems. The movement emerged in the 1970s in the Netherlands, and involved the setting up of 'science shops': physical or metaphorical spaces, often located in and run by traditional research organisations such as universities, that acted as an interface between citizens and researchers.[58] Science shops still exist today, largely in Europe but also in some other countries. Individuals, community groups, or civil society organisations can approach a science shop to ask for assistance with a problem: they might be concerned about unacknowledged environmental degradation, want to know the state of the art in a particular field to help their decision making, or need help evaluating their activities.[59] Science shop employees or volunteers then act as mediators to the wider university, or assist with research themselves. Science shops have often resulted in student research projects focusing on the problems and questions presented, but may also involve desk-based research such as literature reviews or the concerns of citizens being integrated into academics' larger scale projects or programmes. For their enthusiasts, they may even offer a 'new social contract of science and society'.[60]

A similarly institutionalised form of participation also emerged in the Netherlands, though some years later. This format – known as Constructive Technology Assessment, or CTA – was inspired by technology assessment efforts in the United States (in particular). The 1970s had seen the creation of an 'Office of Technology Assessment' in the US Congress, tasked with exploring the potential implications of new technologies, and with assessing their impacts, whether beneficial or adverse.[61] Though by the time of its closure in the 1990s it was 'the leading U.S. evaluator of policy choices concerning technological development',[62] and though it had inspired similar efforts within the policy systems of other countries, it had been criticised for its technocratic approach. It was not open enough to public participation (these critiques went), but focused on technical assessments of risks and benefits, rather than the relation of technologies to public values and priorities.[63] CTA was one response to such criticisms, developing the notion of technology assessment such that it didn't only involve analysis of the possible impacts of already-existing (or almost-ready) technologies by scientists and other experts, but intervened into the innovation process itself and engaged much broader groups. It was a 'design practice', write Johan Schot and Arie Rip, 'in which impacts are anticipated, users and other impacted communities are involved from the start and in an interactive way, and which contains an element of societal learning'.[64] CTA is thus a formal process through which early-stage technologies can be discussed and guided by citizens and other stakeholders. As such it has tended to be institutionalised within governmental or semi-governmental organisations, which are able to have some influence on policy or on research funding. Many states have offices of technology assessment (whether using CTA or more traditional methods): despite its name, for instance, the European Parliament Technology Assessment Network has 25 members from all around the world.[65] Such offices use Technology Assessment methods to engage both experts and citizens, and produce reports and policy guidance on emerging technologies.

Box 2.8: Selling science

Activities such as constructive technology assessment are generally publicly funded, and involve university researchers or others involved in the public research system. However, such scholarship only comprises a fraction of research and development in technoscience: some areas, such as medical or military research or the development of digital tools, involve huge industries with no public remit. This interconnection of science with capitalism was one of the things that did, and does, concern radical science groups, with the pharmaceutical industry one example of these interconnections. In describing the way in which marketing and research are fused in the activities of drug companies, Sergio Sismondo demonstrates the entanglement of capitalism and medicine. Such companies fundamentally need to 'increase their markets'; to do so, 'Pharma companies ghost-manage the production of medical research, they shepherd the key opinion leaders (KOLs) who disseminate the research as both authors and speakers, and finally they orchestrate the delivery of continuing medical education

(CME) courses'.[66] There is thus a significant commercial infrastructure underpinning medical research and the delivery of medical care in North America and beyond, one that is shaping knowledge production and taking it in directions that serve industry interests. As Sismondo writes, '[i]ndividual companies with stakes in specific medical topics can influence knowledge so that their preferred science becomes dominant. ... The flood of knowledge that companies create and distribute is not designed for broad human benefit, but to increase profits'.[67]

Technoscience in deliberative democracy

CTA, science shops and the radical science movement all represent efforts to align research with the needs, priorities and values of publics. In Europe and the United States, in particular, the 21st century saw an intensification of these efforts, and specifically the idea that science needed to be 'democratised'. This has been expressed in different ways: that there is a need for 'extended peer review' of scientific research by affected stakeholders and citizens; that research needs to draw on the substantive knowledges of different publics; that both science and society are changing to become more oriented to specific missions or goals and the creation not just of knowledge, but 'socially robust knowledge'.[68] The overall aim has been a 'richer, wider, and more vibrant empowering of human agency in the deliberate social choice of technological futures', to quote Andy Stirling.[69] As I discuss in Chapters 6 and 8, these moves have in part been triggered by unforeseen impacts of technoscience such as environmental harm or high profile nuclear disasters. In the light of such crises, '[t]he problem we urgently face', writes Sheila Jasanoff, 'is how to live democratically and at peace with the knowledge that our societies are inevitably "at risk". Critically important questions of risk management cannot be addressed by technical experts with conventional tools of prediction'.[70]

These arguments have mingled with discussions from political science concerning the nature of democratic society, and the role of technoscience in it. Democracies take as their starting point the idea that all citizens should be involved in governing a state (the term comes from the Greek word 'Demos', meaning the people). In many democratic systems the involvement of citizens takes place through processes of representation: citizens vote, and the politicians who are elected act on their behalf. Some places also have forms of direct democracy, where citizens vote not just on who should represent them, but to give their answers to specific questions (referenda are one example of this). Increasingly, however, 20th-century political science saw discussion of the idea of *deliberation* as an important supplement to, and maybe even replacement of, established forms of representative democracy.

'Deliberative democratic theory', writes Simone Chambers, 'is a normative theory that suggests ways in which we can enhance democracy. ... Talk-centric democratic theory replaces voting-centric democratic theory'.[71] Deliberative democratic theory argues for the value of deliberation – discussing particular questions – over that of simply voting on them. Rather than assuming that citizens have pre-existing, fixed positions on policy issues (which they can express through voting, either for people or on particular questions), deliberative theory claims that the best decisions are made when people with different experiences and views come together to discuss what the right outcome might be. Their views will shift through such deliberation; perhaps they will, eventually, reach consensus despite their differences. As Smith and colleagues write: 'Hearing the voices of those who are rarely listened to can radically change accepted opinions about what needs to be done. Diversity results in better decision-making.'[72]

Ideas about deliberative democracy have therefore led to a range of efforts to incorporate public deliberation into politics, in the form of everything from participatory budgeting to citizen assemblies and juries on particular issues.* Science policy has been one area to have been affected by this: scholars of deliberation such as John Dryzek[73] have explored how participatory deliberative processes can be used to help make political choices about technoscientific issues. But the rise of deliberative theory has also had a broader impact beyond prompting the use of 'mini-public' formats in the context of science (discussed at more length in Chapter 8); more generally, it has nurtured a sense within many political systems that allowing citizens to participate in policy making is a good thing. An expectation starts to emerge that citizenship is *active*: we should be consulted, allowed to share our views, and have our voices heard in the political choices that may shape our lives. And this is as true of technoscience – as something that indeed comes to impact upon our lives and shape the choices and possibilities available to us – as it is of any other aspect of contemporary societies.

Conclusion

This chapter has set the scene for the rest of the book by exploring histories and imaginations of science–society relations, as well as some of the ways that

* We will spend more time with ideas about deliberation and democracy in Chapter 8, but at this stage it is worth pointing out that these ideas have been subject to criticism along several lines. One of these has involved questioning whether 'democracy' is indeed the universal, straightforward category it is sometimes presented as. In this view theories and practices of deliberative democracy are at best overly simplistic, and at worst a kind of colonisation whereby local principles (of what democracy should look like) are exported to diverse contexts.[h]

technoscience has been related to the public good. Despite the prevalence of languages and metaphors that emphasise the separation of science and society – gaps and ivory towers, for example – its central argument has been that these domains cannot be understood as straightforwardly or consistently distinct. Science and technology are shaped by human choices, values, assumptions, and cultures, while society is constantly impacted by technoscientific developments. This has been true throughout the history of western science (for example as scientists have sought legitimation from public witnesses or funding from public patrons) and continues to be so today (whether through the values that come to be embedded into research or the ways in which it is expected to serve the public good). While there have been a variety of efforts to think about science's role in and responsibility to society, from the notion of a social contract to the radical science movement, recent decades have seen an intensification of efforts to align research with public values and needs. In the context of such discussions technoscience is framed as something that should be subject to deliberation about its goals and outputs, and citizenship as something that is actively concerned with science and technology as much as with other aspects of society. 'Scientific citizenship' is not about passively receiving what those working in technoscience develop, but critically reflecting on and engaging with those developments.

The next chapter picks up on these ideas, in particular by exploring both some concrete ways in which technoscience and society come to shape each other and how citizens debate and intervene in these dynamics. What does the mutual shaping of technoscience and society look like in practice?

References

[a] See Selin, H. (2008). *Encyclopaedia of the History of Science, Technology, and Medicine in Non-Western Cultures*. Springer Science & Business Media.

[b] Alberti, Samuel J.M.M. (2001). Amateurs and Professionals in One County: Biology and Natural History in Late Victorian Yorkshire. *Journal of the History of Biology*, 34, 115–147.

[c] Global Indigenous Data Alliance (nd). *CARE Principles for Indigenous Data Governance*. https://www.gida-global.org/care

[d] Specific texts that discuss this include Collins, H.M. and Pinch, T. (1998). *The Golem: What You Should Know about Science*. Cambridge University Press. Shapin (2010).

[e] See Law, J. (2004). *After Method: Mess in Social Science Research*. Routledge.

[f] See, for instance, Nerlich, B. (2007). Media, Metaphors and Modelling: How the UK Newspapers Reported the Epidemiological Modelling Controversy during the 2001 Foot and Mouth Outbreak. *Science, Technology & Human Values*, 32(4), 432–457. Nerlich, B., Elliott,

R., and Larson, B. (eds) (2009). *Communicating Biological Sciences: Ethical and Metaphorical Dimensions*. Ashgate.

[g] Irwin, A. and Michael, M. (2003). *Science, Social Theory and Public Knowledge*. Open University Press. Jasanoff, S. (2004). *States of Knowledge: The Co-Production of Science and the Social Order*. Routledge.

[h] See Banerjee, S.B. (2022). Decolonizing Deliberative Democracy: Perspectives from Below. *Journal of Business Ethics*, 181(2), 283–299.

[1] Shapin, S. (2012). The Ivory Tower: The History of a Figure of Speech and Its Cultural Uses. *The British Journal for the History of Science*, 45(1), 1–27, at p 14.

[2] Hannam, J. (2011). Explaining the World: Communicating Science through the Ages, in Bennett, D.J., Jennings, R.C., and Bodner, W. (eds) *Successful Science Communication*. Cambridge University Press, pp 31–44, at p 36. http://ebooks.cambridge.org/ref/id/CBO9780511760228A011.

[3] Shapin, S. (2010). *Never Pure: Historical Studies of Science as If It Was Produced by People with Bodies, Situated in Time, Space, Culture, and Society, and Struggling for Credibility and Authority*. Johns Hopkins University Press.

[4] Shapin, S. (1990). Science and the Public, in Olby, R.C., Cantor, G.N., Christie, J.R.R., and Hodge, M.J.S. (eds) *Companion to the History of Modern Science*. Routledge, pp 990–1007, at p 996.

[5] Fan, F. (2012). The Global Turn in the History of Science. *East Asian Science, Technology and Society: An International Journal*, 6(2), 249–258, at p 251.

[6] Barbosa, R.G. (2022). The Origins of Scientific Disciplines: A Counter-history of Western Science. *Indian Journal of History of Science*, 57(3), 202–210.

[7] Hannam (2011).

[8] You can view the painting at https://www.nationalgallery.org.uk/paintings/joseph-wright-of-derby-an-experiment-on-a-bird-in-the-air-pump

[9] Huang, H.-F. (2016). When Urania Meets Terpsichore: A Theatrical Turn for Astronomy Lectures in Early Nineteenth-Century Britain. *History of Science*, 54(1), 45–70. Schaffer, S. (1983). Natural Philosophy and Public Spectacle in the Eighteenth Century. *History of Science*, 21(1), 1–43.

[10] Shapin, S. (1989). The Invisible Technician. *American Scientist*, 77(6), 554–563.

[11] Tansey, E.M. (2008). Keeping the Culture Alive: The Laboratory Technician in Mid-Twentieth-Century British Medical Research. *Notes and Records of the Royal Society*, 62(1), 77–95, at p 78.

[12] Tansey (2008), p 91.

[13] Doing, P. (2004). '"Lab Hands" and the "Scarlet O": Epistemic Politics and (Scientific) Labor'. *Social Studies of Science*, 34(3), 299–323.

[14] Sample, I. (2006, 10 March). Who Really Made Dolly? Tale of British Triumph Descends into Scientists' Squabble. *The Guardian*. https://www.theguardian.com/science/2006/mar/11/genetics.highereducation1

[15] McLaren, C. and Dent, A. (2021). Quantifying the Contributions Technicians Make to Research. *Research Evaluation*, 30(1), 51–56. *Nature*. (2006). Credit Where Credit's Due. *Nature*, 440(7084), 591–592.

[16] Seth, S. (2009). Putting Knowledge in Its Place: Science, Colonialism, and the Postcolonial. *Postcolonial Studies*, 12(4), 373–388.

[17] Seth (2009), p 374.

[18] Das, S. and Lowe, M. (2018). Nature Read in Black and White: Decolonial Approaches to Interpreting Natural History. *Journal of Natural Science Collections*, 6, 4–14, at pp 6–7. See also https://www.nhm.ac.uk/discover/are-natural-history-museums-inherently-racist.html

[19] Bulstrode, J. (2023). Black Metallurgists and the Making of the Industrial Revolution. *History and Technology*, 39(1), 1–41, at p 19.

[20] Bulstrode (2023), p 20.

[21] Hayden, C. (2004). *When Nature Goes Public: The Making and Unmaking of Bioprospecting in Mexico*. Princeton University Press.

[22] Couldry, N. and Mejias, U.A. (2019). Data Colonialism: Rethinking Big Data's Relation to the Contemporary Subject. *Television & New Media*, 20(4), 336–349. Crawford, K. (2021). *Atlas of AI: Power, Politics, and the Planetary Costs of Artificial Intelligence*. Yale University Press.

[23] Bok, K., Sitar, S., Graham, B.S., and Mascola, J.R. (2021). Accelerated COVID-19 Vaccine Development: Milestones, Lessons, and Prospects. *Immunity*, 54(8), 1636–1651.

[24] Martin, E. (1991). The Egg and the Sperm: How Science Has Constructed a Romance Based on Stereotypical Male-Female Roles. *Signs: Journal of Women in Culture and Society*, 16(3), 485–501.

[25] Martin (1991), p 489.

[26] Aarden, E. (2022). Ignorance and the Paradoxes of Evidence-based Global Health: The Case of Mortality Statistics in India's Million Death Study. *Science as Culture*, 31(4), 433–454. Ledebur, S. (2021). Evidence of Undercounting: Collecting Data on Mental Illness in Germany (c. 1825–1925). *Science in Context*, 34(4), 459–478. Stolpe, S., Kowall, B., and Stang, A. (2021). Decline of Coronary Heart Disease Mortality is Strongly Effected by Changing Patterns of Underlying Causes of Death: An Analysis of Mortality Data from 27 Countries of the WHO European Region 2000 and 2013. *European Journal of Epidemiology*, 36(1), 57–68.

[27] Restivo, S. (2017). *Sociology, Science, and the End of Philosophy*. Palgrave Macmillan, pp 253–281.

[28] Moran-Thomas, A. (2020, 5 August). How a Popular Medical Device Encodes Racial Bias. *Boston Review*. https://www.bostonreview.net/articles/amy-moran-thomas-pulse-oximeter/

[29] Buvinic, M. and Levine, R. (2016). Closing the Gender Data Gap. *Significance*, 13(2), 34–37. D'Ignazio, C. and Klein, L.F. (2020). *Data Feminism*. The MIT Press.

[30] Perez, C.C. (2019). *Invisible Women*. Random House. Saini, A. (2017). *Inferior: How Science Got Women Wrong – and the New Research That's Rewriting the Story*. Beacon Press.

[31] Benson, E., Brigg, M., Hu, K., Maddison, S., Makras, A., Moodie, N., and Strakosch, E. (2023). Mapping the Spatial Politics of Australian Settler Colonialism. *Political Geography*, 102, 102855. https://doi.org/10.1016/j.polgeo.2023.102855. Rose-Redwood, R., Blu Barnd, N., Lucchesi, A.H., Dias, S., and Patrick, W. (2020). Decolonizing the Map: Recentering Indigenous Mappings. *Cartographica: The International Journal for Geographic Information and Geovisualization*, 55(3), 151–162.

[32] Boucquey, N., Martin, K.St., Fairbanks, L., Campbell, L.M., and Wise, S. (2019). Ocean Data Portals: Performing a New Infrastructure for Ocean Governance. *Environment and Planning D: Society and Space*, 37(3), 484–503.

[33] See the following volume for one extended account of these dynamics: Bier, J. (2017). *Mapping Israel, Mapping Palestine: How Occupied Landscapes Shape Scientific Knowledge*. MIT Press.

[34] Erickson, I. and Wajcman, J. (2023). Optimizing Temporal Capital: How Big Tech Imagines Time as Auditable. *American Behavioral Scientist*, 67(14), 1755–1770, at p 1757.

[35] Erickson and Wajcman (2023), p 1765.

[36] Erickson and Wajcman (2023), p 1765.

[37] Erickson and Wajcman (2023).

[38] Secher, K. (2013, 26 March). ESOF 2014: Bridging the Gap between Science and Society. *ScienceNordic*. https://sciencenordic.com/communication-denmark-esof2014/esof-2014-bridging-the-gap-between-science-and-society/1383902

[39] Katsenevas, S., Razzano, M., Coyle, P., Angelidakis, S., Marteau, J., and Chaniotakis, E. (2020). Bridging the Gap Between Science and Society Through Citizen Science. [Webinar]. Reinforce. https://www.reinforceeu.eu/events/webinars/bridging-gap-between-science-and-society-through-citizen-science

[40] Jiang, K. (2020, 16 September). Science and Citizenship: Medical School Students Spend Summer Helping Massachusetts State Legislators Craft Science-based Policy. *The Harvard Gazette*. https://news.harvard.edu/gazette/story/ 2020/09/students-work-to-narrow-the-gap-between-scientists-and-society

[41] Secher (2013).

[42] Bensaude-Vincent, B. (2001). A Genealogy of the Increasing Gap between Science and the Public. *Public Understanding of Science*, 10(1), 99–113, at p 107.

[43] Widmalm, S. (2007). Introduction: Science and the Creation of Value. *Minerva*, 45(2), 115–120.

[44] Whyte, K.P. (2013). On the Role of Traditional Ecological Knowledge as a Collaborative Concept: A Philosophical Study. *Ecological Processes*, 2(1), 7, at p 5. https://doi.org/10.1186/2192-1709-2-7
[45] Wyatt, S. (2004). Danger! Metaphors at Work in Economics, Geophysiology, and the Internet. *Science, Technology, & Human Values*, 29(2), 242–261, at p 257.
[46] Shapin (1990), pp 1002–1003.
[47] Martin, B.R. (2003). The Changing Social Contract for Science and the Evolution of the University, in Geuna, A., Salter, A.J. and Steinmueller, W.E. (eds) *Science and Innovation*. Edward Elgar, pp 7–29.
[48] Guston, D.H. and Keniston, K. (1994). *The Fragile Contract: University Science and the Federal Government*. MIT Press.
[49] Shapin (1990).
[50] Mejlgaard, N. and Aagaard, K. (2017). The Social Contract of Science, in Shin, J.C. and Teixeira, P. (eds) *Encyclopedia of International Higher Education Systems and Institutions*. Springer Netherlands, pp 1–4.
[51] Bell, A. (2015, 27 January). Science for the People! *Mosaic Science*. https://web.archive.org/web/20230216133738/https://mosaicscience.com/story/science-people. Bell, A. (2013, 18 July). Beneath the White Coat: The Radical Science Movement. *The Guardian*. https://www.theguardian.com/science/political-science/2013/jul/18/beneath-white-coat-radical-science-movement
[52] Science For The People (nd). *About SFTP*. https://scienceforthepeople.org/about-sftp
[53] Technical Assistance Program (1970, August). *Science for the People*. http://science-for-the-people.org/wp-content/uploads/2015/07/SftPv2n2s.pdf
[54] Bell (2015).
[55] Rose, H. and Rose, S. (1970). *Science and Society*. Penguin.
[56] Rappert, B., Balmer, B. and Stone, J. (2008). Science, Technology, and the Military: Priorities, Preoccupations, and Possibilities in Hackett, E.J., Amsterdamska, O., Lynch, M.E. and Wajcman, J. (eds). *The Handbook of Science and Technology Studies*, MIT Press, pp 719–739. Smit, W.A. (2001). Science, Technology, and the Military, in Smelser, N.J. and Baltes, P.B. (eds) *International Encyclopedia of the Social & Behavioral Sciences*. Elsevier, pp 13698–13704.
[57] Gregory, J. (2022). Personal communication.
[58] Leydesdorff, L. and Ward, J. (2005). Science Shops: A Kaleidoscope of Science-Society Collaborations in Europe. *Public Understanding of Science*, 14(4), 353–372.
[59] SciShops (nd). *Case Studies*. https://www.scishops.eu/resources/case-studies
[60] Leydesdorff and Ward (2005).
[61] Herdman, R.C. and Jensen, J.E. (1997). The OTA Story: The Agency Perspective. *Technological Forecasting and Social Change*, 54(2–3), 131–143.

[62] Bimber, B. and Guston, D.H. (1997). Introduction: The End of OTA and the Future of Technology Assessment. *Technological Forecasting and Social Change*, 54(2–3), 125–130, at p 125.

[63] Bereano, P.L. (1997). Reflections of a Participant-Observer. *Technological Forecasting and Social Change*, 54(2–3), 163–175.

[64] Schot, J. and Rip, A. (1997). The Past and Future of Constructive Technology Assessment. *Technological Forecasting and Social Change*, 54(2), 251–268, at p 255.

[65] EPTA (nd). *Home*. https://eptanetwork.org

[66] Sismondo, S. (2018). *Ghost-Managed Medicine: Big Pharma's Invisible Hands*. Mattering Press, p 11.

[67] Sismondo (2018), p 13.

[68] See Fiorino, D.J. (1990). Citizen Participation and Environmental Risk: A Survey of Institutional Mechanisms. *Science Technology Human Values*, 15(2), 226–243. Funtowicz, S.O. and Ravetz, J.R. (1993). Science for the Post-normal Age. *Futures*, 25(7), 739–755. Nowotny, H., Scott, P., and Gibbons, M. (2001). *Re-Thinking Science: Knowledge and the Public in an Age of Uncertainty*. Polity.

[69] Stirling, A. (2008). 'Opening Up' and 'Closing Down': Power, Participation, and Pluralism in the Social Appraisal of Technology. *Science, Technology & Human Values*, 33(2), 262–294, at p 286.

[70] Jasanoff, S. (2003). Technologies of Humility: Citizen Participation in Governing Science. *Minerva*, 41, 223–244, at p 224.

[71] Chambers, S. (2003). Deliberative Democratic Theory. *Annual Review of Political Science*, 6(1), 307–326, at p 308.

[72] Smith, G., Hughes, T., Adams, L., and Obijiaku, C. (eds) (2021). *Democracy in a Pandemic: Participation in Response to Crisis*. University of Westminster Press, p 4.

[73] Dryzek, J.S., Goodin, R.E., Tucker, A., and Reber, B. (2009). Promethean Elites Encounter Precautionary Publics: The Case of GM Foods. *Science, Technology & Human Values*, 34(3), 263–288.

3

The Mutual Shaping of Technoscience and Society

This chapter is concerned with concrete examples of how the mutual shaping of science, technology, and the social become visible. The ways in which technologies are developed and used is one central instance of this: let's start, then, with three examples of such development.

The first is recent, dating from the time of writing, in 2023. A user on the social media platform Reddit has prompted an artificial intelligence (AI)-based image generator to create a series of selfies of soldiers from throughout history, from Samurai warriors to French soldiers during the First World War.[1] As Jenka Gurfinkel notes in commenting on the images, the result is deeply uncanny for a number of reasons, one of which is the identical way in which each group is depicted as grinning for the camera. Quite aside from the question of whether the battlefield would be a place for smiling selfies, all of the soldiers sport what Gurfinkel describes as the 'American smile': the tooth-revealing grin that results when you are asked to say 'cheese'. These smiles are a result of the training data the AI system was built on, and its reliance on images from the Anglophone internet, where smiling in this way is common. The images are, however, particularly weird if you come from a culture where smiling is done differently, or less frequently. The ways in which emotions are shown are highly specific; to Gurfinkel, as an emigré from Russia to the United States, it grated that all of the different ethnic groups and cultures represented had the same kind of smile. 'AI dominated by American-influenced image sources', Gurfinkel writes, 'is producing a new visual monoculture of facial expressions'. Particular cultural norms – in this case around what smiling should look like and how emotions are expressed – thus come to shape a technology (such as AI image generators), without acknowledgement that local values are being implicitly universalised.*

The second example is older, and concerns the way in which particular technologies may be transported and reinterpreted in new locations. Historian Jean Gelman Taylor tells the story of how Singer sewing machines, after initially being developed in the United States in the 1800s, travelled to the 'Dutch East Indies' – occupied Indonesia – in the early 20th century.[2] While the technology initially formed part of colonial households, acting as a means of demonstrating their modernity, it was rapidly absorbed by local populations and taken up in new ways. Professional tailors made use of the machines to boost production of traditional dress, while individuals could use them to enhance their wardrobes so that they had clothes for different occasions – home, work, religious events – and to create household items that signalled their upward mobility (such as 'hemmed doilies'). As 'new objects [such as sewing machines] reached the Indies', Taylor writes, 'Indonesians became users of them, worked them into their own culture and self-perception as individual human beings'.[3] The sewing machine became something that was integrated into indigenous Indonesian cultures, an object that transcended its initial framing as a western technology designed for the domestic sphere. The technology was appropriated to meet the needs and interests of Indonesian users, in a manner typical of the technologies that moved into colonised sites.†

My final example is again more recent, and concerns the development of 'smart city' technologies. Pouya Sepehr describes how, in Vienna, a team of scientists worked to prototype 'intelligent' pedestrian crossings where, instead of pedestrians having to press a button that would initiate a stop in traffic and a green light signalling they could cross, the system would automatically detect when a pedestrian approached the crossing, and start the process leading to a green light.[4] The aim was to avoid the problem that pedestrians would jaywalk after pressing the button at the crossing, leaving drivers stopped behind an unneccessary red light. While the system

* These kinds of issues with generative AI – with regard to the production of both images and text – are now well established, and relate to diverse forms of bias. I was recently at a conference where a speaker showed an image generated by the prompt 'female president of a surgical society opening the annual congress': the image depicted a scantily clad woman with a stethoscope standing in front of an auditorium of men in suits.

† Taylor is concerned with the sewing machine as one example of how mundane technologies are remade, become the subject of local ingenuity, in diverse contexts. This effect was heightened during the colonial era, which put in place global flows of goods and people, but it is certainly not the case that innovations themselves only emerged from western nations and were used in colonised sites. The kinds of transfers and developments that took place were always more complex.[a]

used state-of-the-art algorithmic technology to recognise the approach of a pedestrian intending to cross (as opposed to just walking by), putting the prototypes into use in the city threw up some problems. It emerged that the developers had forgotten to account for visually impaired pedestrians, who needed different conditions for crossing (including audio signals), and for whom the new system simply didn't work. Furthermore, the city's traffic system had rules regarding how quickly a stop in traffic could be initiated: a traffic light shouldn't change too quickly, and drivers should not have to wait too long. Even though the algorithmic recognition of approaching pedestrians was extremely quick, the duration of the stop sequence meant that pedestrians were waiting for around a minute before they were allowed to cross – a wait that meant that they were still tempted to jaywalk. As Sepehr notes, the saga of the intelligent traffic lights is the latest in a long tradition of urban technologies being used to try and guide the behaviour of people using cities. Streets have been designed to prevent rioting, discourage demonstrations, or render a space inhospitable to the unhoused, for example.[5] In this case design processes (combined with existing traffic infrastructure) simply ignored one type of pedestrian – those with visual impairments – and prioritised the flow of traffic over the possibility of rapid movement for those on foot. Though all of this was (at the time of writing) only realised as a prototype, the case shows how our movements – as pedestrians or as drivers – are defined through the implementation of technologies. They tell us what to do (in this case directly, in the form of red or green lights), and we do it (though not with total obedience, as the continuing presence of jaywalkers indicates).

Generative AI, sewing machines, intelligent traffic lights. These are very different technologies, and their stories foreground different aspects of the role and impacts of technoscience in society. However, in diverse ways they all highlight the constant interaction, the push and pull, between technologies and the social processes that go into designing, implementing, and using them. Generative AI that is built upon a dataset that prioritises particular forms of cultural expression results in a disturbing universalisation of local norms. Sewing machines (and other mundane technologies) are constantly appropriated by their users, pulled into local cultures and used in ways that their designers may not have expected. And traffic lights and urban technologies shape how we can move around urban spaces, defining (for instance) who is welcome in particular areas or how we should behave within them. It is these kinds of dynamics that we will explore in this chapter, looking at the ways in which science and technology are constituted through values, choices, and politics, on the one hand, and how our lives are themselves shaped through the implementation of technologies and scientific knowledge, on the other. If the central question that guides this discussion could be framed as 'Do technologies shape us, or do we shape them?', the answer this chapter gives is: yes – both of those things.

Three myths about technoscience and society

In starting to think about how technoscience and society are co-constituted, or mutually shaped, we can begin by rejecting three myths (or taken-for-granted assumptions) about the relationship between science, technology, and society.

The first is that *science and technology are separate* – that there is a clear distinction between knowledge production, on the one hand, and the utilisation of that knowledge to create tools, innovations, and applications, on the other. This is an idea that circulates in much public discourse, but that is perhaps particularly well exemplified in the motto of the 1933 Chicago World's Fair (which we also met in the previous chapter): 'Science Finds, Industry Applies, Man Conforms'.[6] In this view scientists discover facts about the world, applied researchers in industry use this knowledge to create innovations or products, and humankind adapts to these (the extent to which this latter point is true is something that we will return to). But this tidy distinction does not stand up well to research into technology development, or indeed into scientific practices. Engineers and others in applied domains carry out research and knowledge production, while scientists invent, tinker, and create tools and devices. This point has been particularly vehemently made by historians of science and technology, who have, in studying past inventions and early research, observed the constant intermingling of scientific, technical, and engineering work. As Thomas P. Hughes, a historian of large technical systems such as electrification, writes, '[t]echnology and science, pure and applied, internal and external, and technical and social, are some of the dichotomies foreign to the integrating inventors, engineers, and managers of the system- and networking-building era'.[7] Of course, not all technoscientific practices are the same, and they may be oriented to diverse goals (developing drugs for a particular medical condition or understanding the functioning of that condition, for instance). But it is impossible to draw a sharp line between science (or basic research) and technological development. This is why in this book I use the term technoscience, to acknowledge the continuities and commonalities between research and the development of technologies.*

* The term comes from Actor Network Theory, a body of work I will not discuss in any depth here but which makes the point that any scientific or technical object or practice is networked, connected to many other entities.[b]

Box 3.1: Who gets to be an 'inventor'?

A further critique of myths of technology development is that they focus too much on highly visible, specific inventions, such as the light bulb or the world wide web. In practice, most innovation is iterative, involving small improvements to existing technologies or transferring techniques or tools across to new domains. This focus also means that attention is primarily on specific individuals (framed as heroic inventors), to the detriment of collective labour and the inventive work of those who are readily rendered invisible, such as women or minoritised people. Recent scholarship has started to explore the ingenuity of such individuals, for instance by showing how enslaved peoples in the United States brought their own technologies from their homelands as well as developing their own innovations,[8] or how women created clothing that allowed them greater freedom.[9] Shobita Parthasarathy has discussed the recent example of menstrual hygiene management innovation in India.[10] While one male inventor, Arunachalam Muruganantham, has been celebrated as a case study of 'inclusive innovation' by developing a machine to produce low cost menstrual pads, Parthasarathy argues that discussions around such innovation have ignored the experiences and knowledges of girls and women, including ancient Ayurvedic and other indigenous approaches to managing menstruation. It is easier to focus on singular 'inventors' of technologies that are easily marketised than to investigate existing practices, which requires 'far greater investments of time and resources'.[11]

A second myth is related, and refers to a *linear model of science and technology* in which funding of basic research will lead to applications, commercial products, and societal benefits. This linear model (again something that circulates through public and policy discussion, and that is taken for granted by many funders) is summarised by Benoît Godin as: 'Basic research → Applied research → Development → (Production and) Diffusion'.[12] More generally this model fits into the idea of a social contract between science and society, in which science, if funded from the public purse, will deliver the 'golden eggs' of technologies, innovations, economic growth, and social progress.[13] This imagination guided much science policy over the course of the 20th century and has, Godin argues, become locked into research systems by the use of statistics that classify research activities into different types ('basic', 'applied', 'development', and so on).[14] But the linear model has been subject to critique almost from the moment that it was first articulated. Historians and innovation scholars provide a long list of its shortcomings: many innovations do not emerge from or make use of scientific knowledge; most do not lead to widely used products or social benefits; engineering cultures are (as noted earlier) themselves sites of knowledge production; if science is drawn upon to help create technologies it is often 'old' knowledge, not that at the cutting edge; engagement between different disciplines and cultures, or between industry and academic research, is often difficult and patchy; and in practice the flow may lead in the opposite direction, when the realisation of particular innovation results in the need for certain forms of scientific research.[15] The practical complexity of how innovations emerge and are taken up means, as David Kline and Nathan Rosenberg note, that 'the effects of innovation are hard to measure' and that 'it is a serious mistake to treat an innovation as if it were a well defined, homogeneous thing that could be identified as entering the economy at a precise date'.[16] The linear model has little utility for analysing or managing innovation processes; instead, various scholars have suggested, it continues to circulate in part because it creates an argument for the protection of public funding for basic research.

A final myth further builds on the assumptions embedded in the first two, and is that, while technologies and applied research might be value-driven or subject to politics, *'pure' or 'basic' research is secluded from wider society and value-free*. Technologies might have politics, but research stands aloof from it. Dismissing the first two claims as myths has implications for this one, as, if there are no clear distinctions between science and technology, and no 'pipeline' of knowledge that is applied to produce innovations, then values will be present within research as much as in spaces of technological development. We have already encountered this idea in Chapter 2, when I discussed the entanglement of 'science' and 'society' and argued that society is always present within science, and science within society. It is worth highlighting it again because in this chapter we will spend a lot of time with technologies. In unpacking the ways in which these embody particular assumptions, values, and imaginations the temptation is to think of this embodiment as specifically relating to technological applications, and not to knowledge production. As we will see, technologies imply particular kinds of users, and encourage particular kinds of behaviours (as we have already seen with regard to urban technologies such as smart traffic lights); it is therefore not hard to see how they integrate assumptions and values concerning how users 'should' be. It can be harder to notice how knowledge production is similarly constituted through choices and judgements, any of which can relate to social, political, or ethical values, but, as we saw in Chapter 2, this is certainly the case.

Box 3.2: Gender bias in technoscience

There are numerous ways in which science and technology come to embody particular biases or assumptions. Gender is often part of this: technoscience may reproduce or exacerbate gender stereotypes, or exclude those with particular kinds of bodies. Technologies can render these dynamics particularly visible – take, for example, the enforced delay of the first all-women spacewalk (in 2019) on the discovery that there were not enough spacesuits available in the correct size, a problem that emphasised that '[t]he tools weren't initially designed with women in mind'.[17] Similarly, Katrine Marçal has written about how gender stereotypes can affect which innovations are taken up, arguing that, in the case of the wheeled suitcase, it took a male 'inventor' of a tool that had been used by women for decades for the technology to be taken seriously.[18] But it is not just technologies that may be gendered. Donna Haraway has argued that scientific fields such as primatology are 'politics by other means' in that they offer a space in which mainstream conceptions of notions such as the family or female sexuality may be reproduced (and contested).[19]

Technological determinism and social construction

The mutual constitution of technosciences and society can be thought of as a push and pull, or constant interaction: society shaping technoscience, but also technoscience shaping society. In this section we will look at some of these dynamics of interaction, not as a set of fixed rules as to how they always take place, but as instances of how they can be articulated. In practice the mutual shaping of technoscience and society will always look different in different contexts.

One way of understanding the push and pull of technoscience and society is through the language of technological determinism, on the one hand, and social construction, on the other. Technological determinism is, as the name suggests, the idea that technology is deterministic: in this view, as Sally Wyatt writes, 'technological change causes or determines social change'.[20] This is expressed in the World Fair motto quoted earlier, where humankind simply 'conforms' to technological change, but also remains a taken-for-granted view in much public discussion of technology, where the assumption is that once technologies or innovations have been developed, society will simply adapt to them. Perhaps a central example is the invention of the printing press. Elizabeth Eisenstein's 1979 book *The Printing Press as an Agent of Change*, for example, argues that the development of the printing press was central both to the Reformation – the nascent Protestant movement that emerged in 16th-century Europe – and to wider cultural change. The technology *determined* social and cultural developments: as one contemporary reviewer of Eisenstein's book wrote, 'in an important sense, print caused the Reformation', as well as other forms of change, including the emergence of modern science.[21] Similarly, Benedict Anderson has written about the ways in which print technologies, and specifically newspapers and news media, allowed for the constitution of nations as 'imagined communities' around the world: 'Print-language', he writes, 'is what invents nationalism', in that it allows the discussion and development of shared values and imaginations of a coherent community.[22] Again the circulation of particular technologies come to shape social and cultural possibilities and imaginations.

Such discussions of technological determinism tend to operate on a grand scale, describing the social effects of particular innovations over the course of centuries. But a version of this thinking is also concerned with how specific artefacts or objects shape social practices. Langdon Winner famously asked whether artefacts have politics, answering in the affirmative: technologies, he writes, are 'forms of life. ... The things we call "technologies" are ways of building order in our world'.[23] He gives examples from urban architecture but also of agricultural technologies that remove the need for skilled labour, or nuclear power stations that demand 'authoritarian management'.[24] Any technology embodies assumptions about its users and its use, and those assumptions are political (in that they impose particular forms of life that might be contestable or undesired). Yolande Strengers, for example, has written about what she calls 'resource man', the imagined user that is embedded within smart home technologies (such as energy monitors that display real-time energy usage). This individual is:

> interested in his [*sic*] own energy data, understands it, and wants to use it to change the way he uses energy. He responds rationally to price signals and makes informed decisions based on up-to-date and detailed data provided about the costs, resource units (kilowatt hours), and impacts (greenhouse gas emissions) of his consumption ... the critical issue is that strategies intended for Resource Man overlook, almost entirely, what people actually do in their homes. We do not see the daily domestic dynamics and routines involved in preparing meals; cleaning the body, clothes, and homes; and making spaces and people comfortable, many of which are still predominantly performed by women.[25]

The user that is imagined by smart energy technologies, Strengers suggests, is modelled after energy utility companies themselves: rational, concerned with good resource management, and constantly engaged with fluctuations in energy supply. The tools that are being developed thus aim to assist someone who wants constant information about their energy usage and will respond to this by strategising their energy behaviours. As Strengers writes, this is not a good reflection of 'what people actually do in their homes', in that it disregards the messiness of domestic life and the fact that few of us operate as rational actors all of the time. In particular it ignores the fact that major energy users tend not to be what she calls the 'resource man of the house', but other inhabitants such as teenagers, children, pets, or those with responsibilities for domestic labour, all of whom may have priorities that are different to those of the imagined resource man. Designing for an imagined user is therefore a political act, one that may function to exclude or ignore certain kinds of people, or to make their lives more uncomfortable (Strengers also notes that energy technologies assume a desired ambient temperature of around 22°C; in practice, there are gender and age differences in what is experienced as comfortable[26]).

So technologies shape social interactions and wider society, both on an immediate level (the traffic lights, the smart home monitors) and over long periods (the printing press, the rise of the automobile). But this dynamic is not, as I have written, uni-directional. While technologies embody particular values, and may 'script' the behaviours of their users,[27] few of us are passive recipients of them. Technologies can also be understood as socially constructed, both with regard to their development (to one historian technological determinism is 'self-evidently untrue: human beings construct machines, not the reverse'[28]) and in their contexts of use. Social factors and dynamics guide the conditions under which innovations emerge, how they are designed, and whether or not they are taken up.[29] Similarly, users choose whether or not to engage with certain technologies, adapt or hack them, and appropriate them, drawing them into new contexts or unexpected uses (as we saw with the sewing machine in Indonesia). To be a user is not only to be subjected to scripted behaviours or assumptions about one's nature (such as being a 'resource man'), but to actively engage with technologies and put them to work in ways that meet our particular needs. Indeed, in many cases users are central to technology development: they may test prototypes, offer feedback on early versions of innovations, make requests to developers, or hack or adapt technologies in ways that are then integrated into their later iterations.[30] In one case, for example, users of an early home computer, the TRS-80, refused to let it die even when its designers and producers stopped making and supporting it in the early 1980s.[31] Christina Lindsay tells the story of how a community developed around the TRS-80, one that involved assisting in its maintenance, developing it further, and publicising it to others (it gained a new lease of life with the rise of emulators in the 1990s, meaning that it could be replicated on other – much newer – computers). In such cases distinctions between designers and users break down entirely: as Lindsay writes, '[j]ust because a technology is no longer being produced or sold does not mean that it is no longer being used or even, as in the case of the TRS-80, developed further'.[32]

Box 3.3: Non-users, non-use, and maintenance

It is not only users of particular technologies who help shape how those technologies are realised in the world, but non-users and others who resist them. Non-participation in particular online spaces (such as social media), for instance, will shape the ways in which such platforms develop, given the way in which they harvest data from their users.[33] Similarly, one study explored resistance to digital technologies in music communities, looking at how analogue devices (such as synthesisers) were framed as more authentic or as sounding better. As the authors note, such non-use feeds back into development of digital music technologies when developers seek to mimic the 'retro' sound or design that is prized by analogue enthusiasts.[34] In the same way, recent scholarship has emphasised that *maintenance* should be considered as important as innovation. 'Maintenance and repair [and] the mundane labour that goes into sustaining functioning and efficient infrastructures, simply has more impact on people's daily lives than the vast majority of technological innovations', write Andrew Russell and Lee Vinsel, in an article that argues the need to celebrate those who do this maintenance work.[35] Maintenance, repair, and non-use thus shape technologies as much as innovation and use.

Box 3.4: What are 'technology' and 'innovation'?

Much of my discussion so far has taken for granted colloquial understandings of technology and of innovation. In this view, 'technologies' are discrete tools that are developed to assist with particular tasks – from nuclear power stations to sewing machines or generative AI – while 'innovation' is the creative activity that leads to such technologies. As we have already seen in discussing the relation between science and technology, however, this is at best a convenient simplification. In particular it is important to note that technological innovation and novelty emerge in many different sites, not just those commonly framed as 'high-tech' (such as digital technology). We also find it in examples such as *jugaad*, 'a colloquial Hindi and Punjabi word meaning an innovative fix or a simple workaround ... used to describe creative and hands-on solutions to making that bend the rules of daily life, enabling practices defined by casual ubiquity'.[36] 'Innovation' is thus a complex and contested notion. While notions such as 'inclusive' or 'grassroots' innovation emphasise that it may emerge from lived experiences of particular needs,[37] in many policy and business contexts 'innovation' has become an almost fetishised notion that is straightforwardly tied to economic development (something we will return to in Chapter 9). In contexts such as life science start-ups, making a *promise* of innovation can be more important than having concrete products.[38]

The mutual shaping of digital technologies and society

The dynamics of society shaping technoscience and technoscience shaping society are perhaps particularly visible in the context of digital technologies.

Sites such as Google or YouTube reach billions globally,[39] while the internet generally and social media particularly are becoming increasingly entangled with the development of AI technologies. How do we see the kinds of interactions and intersections described earlier playing out in this space?

On the one hand there are many claims that these technologies will determine social life and structures. While early suggestions that the internet would usher in a new age of connectivity and democratic engagement have fallen into abeyance,[40] digital technologies are instead framed as bringing about new forms of business and of economic productivity, for instance by enabling a new 'data economy' that will inevitably generate wealth.[41] It is also inarguable that social media, smartphones, and other digital tools and platforms are shaping many aspects of many people's lives, from the rise of the gig economy to new forms of intimacy or possibilities for online community around niche interests.[42] On the other hand, it is clear that users of such technologies are actively engaged in using, recreating, or resisting them, rather than being passive recipients, and that the technologies themselves are not neutral or objective but constituted through particular assumptions about users and the world around them. In particular, there are now plentiful examples of the ways in which digital tools reproduce the biases and limited perspectives of their developers, an effect that is particularly insidious when they are presented as objective and widely applicable. Ruha Benjamin has written about these dynamics as the 'New Jim Code' (a reference to US 'Jim Crow' laws that enforced racial segregation): 'the employment of new technologies that reflect and reproduce existing inequities but that are promoted and perceived as more objective or progressive than the discriminatory systems of a previous era'.[43] As with the example of the 'American smile' discussed at the start of the chapter, the danger is that the behaviours, preferences, or priorities of a limited subset of the world's population are represented as universal, and that inequities are smuggled into seemingly neutral technologies.

Indeed, there have now been numerous high-profile examples of how digital tools designed for general use reproduce the biases of the people and materials involved in creating them. In 2016 the Microsoft chatbot Tay, who had been designed to learn from the social media platform Twitter and those that she interacted with, was rapidly taken offline when she started tweeting sexist, racist, and conspiracy theory-oriented content. 'Tay's learning algorithms replicated the worst racism and sexism of Twitter', write Gina Neff and Peter Nagy.[44] Computer scientist Joy Buolamwini has described how facial recognition systems trained on predominantly light skin meant that such systems were ineffective for people of colour,[45] a problem that led to her creation of the Algorithmic Justice League in an effort to highlight instances where bias was coded into software.[46] Similarly, research has shown that beauty apps and filters – which can be used both for still images and video and 'beautify' one's face – function to reinforce a specific beauty ideal. Such technologies 'homogenize[s] the visual aesthetics of faces … [and] make them conform with a canon of beauty of white people'.[47] Such issues are not confined to online spaces or to images: Dirk Hovy and colleagues have documented how machine learning-based translation tools reproduce the demographic biases of their training data, effectively making translated language sound 'older and more male' than the original input.[48]

Such biases can and are having substantive effects. One key area for AI and machine learning technologies is in welfare systems or other forms of state action, where such tools are often understood as enabling efficiency and cost-saving.[49] In Austria, for example, the Public Employment Service developed an algorithm to assist its staff in making judgements concerning the resources job seekers should be given access to, classifying them with regard to how likely the system assessed them as being to find a job within the next months. While the use of the algorithm was halted by the Data Protection Service (and is, in 2024, still in limbo), it was controversial because it based its recommendations on existing labour statistics (meaning that if you were a woman or immigrant, for example, you automatically received a lower score, because these groups have traditionally found it harder to find work) and thereby reproduced the inequalities of the labour market. Those already subject to marginalisation, discrimination, or minoritisation had these effects compounded by the logic of the algorithm, which directed resources away from 'hopeless' cases.[50] Similarly, the emerging field of predictive policing involves a raft of digital technologies that use data from historic and current crimes in particular areas to make predictions as to where there is a high likelihood of criminal activity, and therefore where to focus law enforcement resources. While these technologies are at different stages, and are used differently in different national contexts, it seems likely that they will involve a similar effect as the Employment Service algorithm, reinforcing existing patterns with regard to inequity in policing practice. As Simon Egbert and Matthias Leese note, such systems can only work with the data that they have, and '[o]nly crime that has been detected or reported becomes part of official crime data and thus part of any analysis'; in the same way, there are known biases with regard to who gets arrested (poorer people, people of colour). The danger is therefore that predictive policing tools will 'reinforce discrimination and stigmatization'.[51]

Box 3.5: Artificial intelligence's contemporary impacts

While much discussion of AI technologies has focused on their potential to create 'existential risks' in the future,[52] they are already having impacts on many people's lives. As well as the use of algorithms and related forms of decision-making assistance in welfare systems described in the main text, facial recognition technologies are increasingly used by the police in some countries. Such systems are, as with any form of AI, only as good as the datasets on which they are trained. 2019 saw the first (known) wrongful arrests based on the use of facial recognition technology, with one man spending ten days in prison before being released. As well as the possibility of such tools replacing other forms of investigation, facial recognition is known to be less accurate for women and people of colour, who therefore have a higher chance of being misidentified.[53] Such negative impacts of AI tend to disproportionately affect those who are already vulnerable, such as the unemployed, or those with criminal records. As Ruha Benjamin has argued, such technologies thus exacerbate inequality while simultaneously being presented as objective.[54]

These issues with digital technologies have now received substantial public attention, and there are corresponding efforts to correct their biases and imperfections. One way to do this is to ensure that the communities who are producing such technologies are more diverse: if developers include people with a wider range of backgrounds and identities than is currently the case in the tech industry, then what Joy Buolamwini calls the 'down the hall test' – asking colleagues in the next door offices for feedback – will involve input from more diverse perspectives, and biases and inadequacies are more likely to be identified.[55] Similarly, there are now more extensive efforts to clean datasets and to render them more representative. At the same time it is important not to view 'good' datasets (and their use as training data for AI and algorithmic technologies) as the sole or even primary answer to the problems inherent to digital tools. Creating more representative or 'better' datasets (for instance by removing offensive content or ensuring that people from a wider range of nationalities or backgrounds are present within them) may have other effects that are similarly undesirable: the need to clean datasets has, for example, resulted in a new form of poorly paid labour where individuals are tasked with removing inappropriate content. In many cases this is outsourced to low-income countries, or carried out via cloud-based platforms, where data workers handle disturbing and at times highly distressing material for minimal compensation. Similarly, the focus on training data may distract from other impacts of AI and digital technologies. Vladan Joler and Kate Crawford's 'Anatomy of an AI System' traces the diverse forms of human labour, planetary resources, and data that go into the creation of a system such as Amazon's Alexa, describing the complex supply chains and diverse materials that are essential to it but that are largely invisible to the end user.[56] AI, they suggest, involves the depletion of rare earth minerals and environmental degradation associated with this; the use (and monetisation) of user data and a corresponding erosion of privacy; climate costs deriving from the huge amounts of energy required for data processing; precarious and at times dangerous human labour in, for instance, assembling electronic devices or clearing toxic waste when such devices are later abandoned; and the reinforcement of global capitalism in a manner that concentrates wealth in the hands of a few individuals such as business owners. It is therefore essential to be attentive to all of these diverse dimensions (and more) when considering the ways that this particular technology is impacting society (and the global environment), and how it is itself being shaped by existing social structures.

Box 3.6: The 'algorithmic colonisation of Africa'

AI is in many ways an extractive technology – it requires huge amounts of energy to support the data processing that goes into machine learning, that data is itself often derived from the work of uncredited authors or other creators, and digital technologies in general rely on scarce natural resources such as rare earth minerals. In its reliance on the extraction of resources and value from diverse global sites and populations, AI development may be reproducing colonial dynamics. Computer scientist Abeba Birhane has written about what she has called the 'algorithmic colonisation of Africa', a process in which a rush to access and monetise data in the Global South disregards actual needs, skills, and priorities. '[T]his discourse of "mining" people for data', she writes, 'is reminiscent of the coloniser attitude that declares humans as raw material free for the taking'.[57] At the same time much of the work of content moderation and dataset cleaning that is essential to the tech industry is outsourced to the Global South, where costs are lower. There are signs, however, that the visibility of this work is increasing: in 2023, 150 African content moderators voted at a Nairobi summit to unionise, allowing for a collective voice in arguing for fairer working conditions.[58]

Encountering and reflecting on technology

Digital technologies – an almost ubiquitous form of technological development in the 2020s – are therefore a central case for observing how technoscience and society are mutually constituted. They illustrate the ways in which technologies incorporate value judgements and assumptions about their contexts of use, as well as how they come to shape our day-to-day lives (the algorithm that denies us access to benefits, the beauty tool that makes us look thinner or whiter). Importantly, they also teach us to look for these dynamics at a variety of scales and timeframes. The use of historic or limited training data certainly highlights the ways in which technologies are modelled after their makers' assumptions about the world, and embody the politics of these, but discussion of AI (in particular) also forces us to acknowledge less visible global impacts of digital technologies. Emerging AI technologies may reproduce the biases and injustices of the material they are trained on – as in the Public Employment Service algorithm or facial recognition tools that cannot recognise darker skin – but they also contribute to the climate crisis, reinforce global capitalism, and normalise a world in which industry takes the lead on technology development, with little regulation or public debate concerning how such technologies emerge into the world. Technoscience is thus entangled with society both at the level of individual users – the ways that we choose to engage or not with particular technologies or forms of knowledge, and how this (dis)engagement in turn comes to shape how they are realised – and as global processes and intersections (for instance in the form of a reinstantiation of colonial logics that extract value from the Global South).

All of this suggests that the ways in which science and technology are developed and put into use deserves careful reflection. While this is an idea that I will discuss in more detail in Chapters 5 and 8, in this final section I want to consider some examples of how different groups are actively choosing to engage with particular technologies. What does it look like to collectively reflect on – and perhaps resist – technologies as they enter our worlds?

One example comes from the work of Jameson M. Wetmore on the ways in which Amish communities (specifically Old Order Amish communities) have traditionally engaged with technology. While the Amish – a North American Protestant religious sect – are often understood as rejecting all technology from the 20th century onwards, Wetmore explains that this is not the case. Instead they believe that 'technologies can reinforce social norms, enable or constrain the ways that people interact with one another, and shape a culture's identity' (as indeed has been the theme of this chapter), and therefore that any technology deserves reflection and debate before its adoption.[59] Importantly, Amish people do not assume that all progress and innovation is necessarily good, an approach that they view as being different to the mainstream culture around them. Amish communities therefore regularly meet to discuss whether new technologies should be collectively adopted, basing their judgement on whether it will further their values. 'Machinery is not wrong in itself', Wetmore quotes one of his interlocutors as saying, 'but if it doesn't help fellowship you shouldn't have it'.[60] Prioritising values such as communal living and humility have led them to reject technologies such as automobiles, because these 'may cause a person to focus on him or herself as an individual and thereby neglect the group'. Any technology is thus carefully considered with regard to what it is likely to do to the community, and whether it is aligned with its shared values, with assessments taking place in collective meetings.

The thoughtful and careful consideration that Amish people give technologies is certainly aided by the degree to which they live and work in small communities that are oriented to a clear set of shared values: most forms of organised deliberation on technoscience do not take place under these conditions, and are correspondingly more complicated (as we will discuss in Chapter 8). But reflection and resistance may also emerge in less centralised ways. Again, digital technologies are a central case here, connected to an increasing sense that such technologies have emerged without sufficient reflection or caution. The use and monetisation of citizens' data in 'surveillence capitalism', the increasing necessity of social media use – and corresponding data 'donation' – in order to participate in society, a lack of privacy as data from devices, cities, and any form of public material is harvested by states or companies:[61] all of these aspects of the current environment (one in which 14 per cent of all internet visits worldwide are to YouTube, and 4 per cent to Facebook[62]) has meant that there have been increasing concerns, both in public discussion and within specific communities, about a 'democratic deficit' around the emergence of digital technologies, and specifically the extent to which they are entangled with large companies such as Amazon, Meta, and Google. Democracy is at risk through the behaviour of such platforms: '[e]xtracting the essence of our humanity in data form and then using it to manipulate our behaviour is as unethical as child labour', said one former Facebook investor in 2021 in the wake of leaked internal documents concerning the site's practices.[63]

While some have responded with apathy or alienation to such concerns, what Helen Kennedy has called 'living with data' is perhaps more complex than simply responding to these developments as straightforwardly positive or negative.[64] As she and colleagues write, '[p]eople's knowledge of and feelings about data uses are not static … they can change in the process of thinking, reading, or talking about them'.[65] In addition, there are now a range of emergent forms of grassroots 'data activism' through which citizens are seeking to intervene in the landscape of digital technology. I have already mentioned the Algorithmic Justice League, a campaigning organisation that carries out a range of activities oriented to raising public awareness of algorithmic bias and finding strategies to combat this: one project, for example, explores how the use of make-up and accessories inspired by drag can disrupt facial recognition technologies.[66] More broadly scholars distinguish between proactive and reactive data activism, where:

> Re-active data activism comprises the practices of resistance to the threats to civil and human rights that derive from corporate and government privacy intrusion. Pro-active data activism embraces those individuals and civil society organizations that take advantage of the possibility for social change and civic engagement offered by big data. Re-active and pro-active data activism … are enabled by software to

manipulate data or to shield one's online interactions from intrusion and automatized collection.[67]

Both reactive and proactive data activism comprise a wide range of activities: for example disengaging from particular platforms or social media, using encryption to enhance privacy, or using alternative infrastructures to those provided by large companies (Mastodon rather than Twitter, for example), on the one hand, and citizen-compiled datasets, data journalism, advocacy for open data practices, or 'civic hacking' (the use of data to try and bring about social change), on the other.[68] In many cases reactive and proactive forms of activism combine. The organisation Tactical Tech, for instance, offers resources for those who might take on a reactive role – such as a 'data detox kit' that helps users to understand the status of their online privacy and security – while also actively advocating for and supporting reflection on desirable digital futures.[69] While, within this landscape, the degree to which such activism is coordinated varies, in all cases there is substantive reflection on the kinds of digital technologies we want to have in our lives, and action taken to try and ensure this. In common with the Amish communities described by Jameson M. Wetmore, data activists explore how particular technologies align with individual and collective values, and seek to find ways of moving towards desirable futures.

Box 3.7: Data activism in practice

What can data activism look like in practice? One example is the Argentinian project #NiUnaMenos (Not One Woman Less), which created a grassroots National Index of Male Violence in a context in which femicide was rife but had received little media or policy attention.[70] In such cases counting what has previously not been considered important or relevant enough to measure is an intervention into public debate, and a critique of the values implied by the standard datasets gathered by states.[71] The COVID-19 pandemic was also a key moment for citizen-generated data and analysis. In 2020 the German collective zerforschung installed sensors on trains on the Berlin U-Bahn system that could communicate with the German contact tracing app, and that were thus able to track how many infections were reported via this. The results meant that they could suggest the 'rule of thumb' that in each full U-Bahn there was an infected person, despite reassurances from the transport authority that it was safe to travel.[72]

Conclusion

This chapter has explored concrete ways in which technoscience and society come to shape each other, looking in particular at (digital) technology development and use. While it is important not to impose clear distinctions between scientific and technological research – in practice these always overlap, while basic research does not lead to technology development in any consistent way – technologies, and especially digital technologies, offer a central example of the ways in which the choices, assumptions, and values of human actors come to shape the nature of technoscience, and to impose particular 'forms of life' on those who use its products. There is a constant push and pull between technoscience and society: technologies shape social life (the idea of technological determinism), but at the same time they are also created, appropriated, given meaning, and repurposed by people (in processes of social construction). What is perhaps most productive is to explore how these dynamics unfold in particular sites or contexts. Digital technologies, in particular, offer an important example of how the biases, assumptions, and pathologies of wider society may be reproduced in technology development, whether that is through reliance on limited datasets that universalise particular forms of experience (US culture or white skin) or capitalism as the key framework for the creation and dissemination of technologies. At the same time, users of technologies are never passive. Even those of us who experience particular technologies as invisible or unproblematic fold them into our day-to-day lives, giving them meaning through this contextualisation, while others may develop more explicit forms of reflection, rejection, or protest around them. Given the ubiquity of digital technologies, data activism is one important example of this, but (as we will also see in Chapter 5) there are a range of histories and traditions of public reflection and activism regarding the development and use of technoscience.

This chapter has therefore further grounded the book's central claim that science, technology, and society are constantly entangled with each other. I have focused on technologies, but these are, of course, just one aspect of the ways in which technoscience becomes visible within everyday life and popular culture. The next chapter looks at another aspect, that of public representations of science and technology (and those involved in them). How do shared imaginations of technoscience come to constitute both the practice of science and experiences of collective life?

References

[a] See discussion in Choi, H. (2017). The Social Construction of Imported Technologies: Reflections on the Social History of Technology in Modern Korea. *Technology and Culture*, 58(4), 905–920.

[b] For one treatment that discusses the overlap between scientific and engineering practices see Latour, B. (1987). *Science in Action: How to Follow Scientists and Engineers through Society*. Harvard University Press.

[1] See the images and discussion at: Gurfinkel, J. (2023, 27 March). AI and the American Smile. *Medium*. https://medium.com/@socialcreature/ai-and-the-american-smile-76d23a0fbfaf

[2] Taylor, J.G. (2012). The Sewing-Machine in Colonial-Era Photographs: A Record from Dutch Indonesia. *Modern Asian Studies*, 46(1), 71–95.

[3] Taylor (2012), p 94.

[4] Sepehr, P. (2024). Mundane Urban Governance and AI Oversight: The Case of Vienna's Intelligent Pedestrian Traffic Lights. *Journal of Urban Technology*, 1–18. https://doi.org/10.1080/10630732.2024.2302280

[5] Winner, L. (2001). *The Whale and the Reactor: A Search for Limits in an Age of High Technology*. University of Chicago Press. Petty, J. (2016). The London Spikes Controversy: Homelessness, Urban Securitisation and the Question of 'Hostile Architecture'. *International Journal for Crime, Justice and Social Democracy*, 5(1), 67–81.

[6] Widmalm, S. (2007). Introduction: Science and the Creation of Value. *Minerva*, 45(2), 115–120.

[7] Hughes, T.P. (1986). The Seamless Web: Technology, Science, Etcetera, Etcetera. *Social Studies of Science*, 16(2), 281–292, at p 286.

[8] James, P. (2021, 23 September). 300 Years of African-American Invention and Innovation. *The MIT Press Reader*. https://thereader.mitpress.mit.edu/300-years-of-african-american-invention-and-innovation/

[9] Jungnickel, K. (2023) Clothing Inventions as Acts of Citizenship? The Politics of Material Participation, Wearable Technologies, and Women Patentees in Late Victorian Britain. *Science, Technology, & Human Values*, 48(1), 9–33.

[10] Parthasarathy, S. (2022). How Sanitary Pads Came to Save the World: Knowing Inclusive Innovation through Science and the Marketplace. *Social Studies of Science*, 52(5), 637–663.

[11] Parthasarathy (2022), p 646.

[12] Godin, B. (2006). The Linear Model of Innovation: The Historical Construction of an Analytical Framework. *Science, Technology, & Human Values*, 31(6), 639–667, at p 639.

[13] Shapin, S. (1990). Science and the Public. In Olby, R.C., Cantor, G.N., Christie, J.R.R., and Hodge, M.J.S. (eds) *Companion to the History of Modern Science*. Routledge, pp 990–1007.

[14] Godin (2006).

[15] Kline, S.J. and Rosenberg, N. (1986). An Overview of Innovation, in National Research Council, *The Positive Sum Strategy: Harnessing Technology for Economic Growth*. The National Academies Press, pp 275–306, at p 275. Rosenberg, N. (1991). Critical Issues in Science Policy Research. *Science and Public Policy*, 18(6), 335–346.

[16] Kline and Rosenberg (1986), p 283.

[17] Wei-Haas, M. (2019, 18 October). First All-woman Space Walk Puts Spotlight on Spacesuit Design. *National Geographic*. https://www.nationalgeographic.com/science/article/first-all-women-spacewalk-suit-design

[18] See Marçal, K. (2021, 24 June). Mystery of the Wheelie Suitcase: How Gender Stereotypes Held Back the History of Invention. *The Guardian*. https://www.theguardian.com/lifeandstyle/2021/jun/24/mystery-of-wheelie-suitcase-how-gender-stereotypes-held-back-history-of-invention. Marçal, K. (2021). *Mother of Invention: How Good Ideas Get Ignored in a World Built for Men*. HarperCollins UK.

[19] Haraway, D.J. (1984). Primatology is Politics by Other Means. *PSA: Proceedings of the Biennial Meeting of the Philosophy of Science Association, 1984*, 489–524.

[20] Wyatt, S. (2008). Technological Determinism Is Dead; Long Live Technological Determinism, in Hackett, E.J., Amsterdamska, O., Lynch, M., and Wajcman, J. (eds) *The Handbook of Science and Technology Studies* (3rd edn). MIT Press, pp 165–180, at p 168.

[21] Kingdon, R.M. (1980). Review of the Printing Press as an Agent of Change: Communications and Cultural Transformations in Early-Modern Europe. *The Library Quarterly: Information, Community, Policy*, 50(1), 139–141, at p 140.

[22] Anderson, B. (2006). *Imagined Communities: Reflections on the Origin and Spread of Nationalism*. Verso, p 134.

[23] Winner (2001), p 28.

[24] Winner (2001), p 176.

[25] Strengers, Y. (2014). Smart Energy in Everyday Life: Are You Designing for Resource Man? *Interactions*, 21(4), 24–31, at pp 26, 28.

[26] Choi, J., Aziz, A., and Loftness, V. (2010). Investigation on the Impacts of Different Genders and Ages on Satisfaction with Thermal Environments in Office Buildings. *Building and Environment*, 45(6), 1529–1535.

[27] For a discussion of how technologies embody 'scripts' that their users are expected to – but do not always – follow, see Akrich, M. (1992). The De-scription of Technical Objects, in Bijker, W.E. and Law, J. (eds) *Shaping Technology/Building Society*. MIT Press, pp 259–263.

[28] Quoted in Dafoe, A. (2015). On Technological Determinism: A Typology, Scope Conditions, and a Mechanism. *Science, Technology, & Human Values*, 40(6), 1047–1076, at p 1057.

[29] See Bijker, W.E. (1997). *Of Bicycles, Bakelites, and Bulbs: Toward a Theory of Sociotechnical Change*. MIT Press.

[30] See Oudshoorn, N. and Pinch, T. (eds) (2003). *How Users Matter: The Co-construction of Users and Technologies*. MIT Press. In addition users may 'domesticate' technologies: see Hirsch, E. and Silverstone, R. (2003). *Consuming Technologies: Media and Information in Domestic Spaces*. Routledge.

[31] Lindsay in Oudshoorn and Pinch (2003).

[32] Lindsay in Oudshoorn and Pinch (2003), p 50.

[33] Casemajor, N., Couture, S., Delfin, M., Goerzen, M., and Delfanti, A. (2015). Non-participation in Digital Media: Toward a Framework of Mediated Political Action. *Media, Culture & Society*, 37(6), 850–866.

[34] Thorén, C. and Kitzmann, A. (2015). Replicants, Imposters and the Real Deal: Issues of Non-use and Technology Resistance in Vintage and Software Instruments. *First Monday*, 20(11). https://doi.org/10.5210/fm.v20i11.6302

[35] Russell, A. and Vinsel, L. (2016). Hail the Maintainers. *Aeon Essays*, 7. https://aeon.co/essays/innovation-is-overvalued-maintenance-often-matters-more

[36] Braybrooke, K. and Jordan, T. (2017). Genealogy, Culture and Technomyth. *Digital Culture & Society*, 3(1), 25–46, at p 30.

[37] Collier, S.J., Cross, J., Redfield, P., and Street, A. (eds) (2017). Little Development Devices/Humanitarian Goods. *limn.* https://limn.it/issues/littledevelopment-devices-humanitarian-goods/. Smith, A., Hargreaves, T., Hielscher, S., Martiskainen, M., and Seyfang, G. (2016). Making the Most of Community Energies: Three Perspectives on Grassroots Innovation. *Environment and Planning A*, 48(2), 407–432.

[38] Birch, K. (2023). Reflexive Expectations in Innovation Financing: An Analysis of Venture Capital as a Mode of Valuation. *Social Studies of Science*, 53(1), 29–48.

[39] Petrosyan, A. (2024, 13 February). Internet Usage Worldwide: Statistics & Facts. *Statista*. https://www.statista.com/topics/1145/internet-usage-worldwide/#topicOverview

[40] Morozov, E. (2011). *The Net Delusion: The Dark Side of Internet Freedom*. PublicAffairs.

[41] See, for instance, Cavanillas, J.M., Curry, E., and Wahlster, W. (eds) (2016). *New Horizons for a Data-Driven Economy*. Springer International Publishing.

[42] Andreassen, R., Petersen, M.N., Harrison, K., and Raun, T. (eds) (2017). *Mediated Intimacies: Connectivities, Relationalities and Proximities*. Routledge.

[43] Benjamin, R. (2019). *Race After Technology: Abolitionist Tools for the New Jim Code*. John Wiley & Sons, p 6.

[44] Neff, G. and Nagy, P. (2016). Talking to Bots: Symbiotic Agency and the Case of Tay. *International Journal of Communication*, 10, 4915–4931.

[45] Buolamwini quoted in Tucker, I. (2017, 28 May). 'A White Mask Worked Better': Why Algorithms are Not Colour Blind. *The Guardian*. https://www.theguardian.com/technology/2017/may/28/joy-buolamwini-when-algorithms-are-racist-facial-recognition-bias

[46] See AJL (nd). *About*. https://www.ajl.org/about. Buolamwini, J. and Gebru, T. (2018). Gender Shades: Intersectional Accuracy Disparities in Commercial Gender Classification. *Proceedings of the 1st Conference on Fairness, Accountability and Transparency*, 77–91. https://proceedings.mlr.press/v81/buolamwini18a.html

[47] Riccio, P. and Oliver, N. (2023). Racial Bias in the Beautyverse: Evaluation of Augmented-Reality Beauty Filters, in Karlinsky, L., Michaeli, T., and Nishino, K. (eds) *Computer Vision – ECCV 2022 Workshops*. Springer Nature Switzerland, pp 714–721, at p 718.

[48] Hovy, D., Bianchi, F., and Fornaciari, T. (2020). 'You Sound Just Like Your Father': Commercial Machine Translation Systems Include Stylistic Biases. *Proceedings of the 58th Annual Meeting of the Association for Computational Linguistics*, pp 1686–1690.

[49] For a detailed treatment see Eubanks, V. (2017). *Automating Inequality: How High-tech Tools Profile, Police, and Punish the Poor* (1st edn). St. Martin's Press.

[50] Allhutter, D., Cech, F., Fischer, F., Grill, G., and Mager, A. (2020). Algorithmic Profiling of Job Seekers in Austria: How Austerity Politics Are Made Effective. *Frontiers in Big Data*, 3(5).

[51] Egbert, S. and Leese, M. (2020). *Criminal Futures: Predictive Policing and Everyday Police Work* (1st edn). Routledge, pp 191–192.

[52] Hanna, A. and Bender, E. (2023, 12 August). AI Causes Real Harm: Let's Focus on That Over the End-of-Humanity Hype. *Scientific American*. https://www.scientificamerican.com/article/we-need-to-focus-on-ais-real-harms-not-imaginary-existential-risks/

[53] Johnson, K. (2022, 7 March). How Wrongful Arrests Based on AI Derailed 3 Men's Lives. *Wired*. https://www.wired.com/story/wrongful-arrests-ai-derailed-3-mens-lives

[54] Benjamin, R. (2019). *Race After Technology: Abolitionist Tools for the New Jim Code*. John Wiley & Sons.

[55] Buolamwini quoted in Tucker (2017).

[56] Crawford, C. and Joler, V. (2018). *Anatomy of an AI System: The Amazon Echo as an Anatomical Map of Human Labor, Data and Planetary Resources*. SHARE Lab, Share Foundation, and The AI Now Institute, NYU. https://anatomyof.ai

[57] Birhane, A. (2020). Algorithmic Colonization of Africa. *SCRIPTed*, 17(2), 389–409, at p 398.

[58] *Foxglove* (2023, 15 May). Looking Back at the Nairobi Content Moderation Summit – Where the Vote to Form the First African Content Moderators' Union Took Place. *Foxglove*. https://www.foxglove.org.uk/2023/05/15/nairobi-content-moderation-summit

[59] Wetmore, J.M. (2007). Amish Technology: Reinforcing Values and Building Community. *IEEE Technology and Society Magazine*, 26(2), 10–21.

[60] Wetmore (2007), pp 11–12.

[61] See Zuboff, S. (2019). Surveillance Capitalism and the Challenge of Collective Action. *New Labor Forum*, 28(1), 10–29.

[62] Bianchi, T. (2024, 24 January). Most Popular Websites Worldwide in April 2023, Based on Share of Visits. *Statista*. https://www.statista.com/statistics/265668/global-top-websites-ranked-by-visit-share

[63] Milmo, D. (2021, 3 November). Democracy at Risk if Facebook Does Not Change, Says Former Zuckerberg Adviser. *The Guardian*. https://www.theguardian.com/technology/2021/nov/03/democracy-at-risk-if-facebook-does-not-change-says-former-zuckerberg-adviser

[64] Kennedy, H. (2018). Living with Data: Aligning Data Studies and Data Activism Through a Focus on Everyday Experiences of Datafication. *Krisis*, 1, 17–30.

[65] Taylor, M., Kennedy, H., and Oman, S. (2024). Challenging Assumptions about the Relationship between Awareness of and Attitudes to Data Uses amongst the UK Public. *The Information Society*, 40(1), 32–53, at p 51.

[66] AJL (nd). *Get Ready to Drag the Cistem!* https://www.ajl.org/drag-vs-ai. See also De Leon, R. (2021, 17 September). Researchers Defeated Advanced Facial Recognition Tech Using Makeup. *Vice*. https://www.vice.com/en/article/k78v9m/researchers-defeated-advanced-facial-recognition-tech-using-makeup

[67] Milan, S. and Gutiérrez, M. (2015). Citizens' Media Meets Big Data: The Emergence of Data Activism. *MEDIACIONES*, 11(14), 120–133, at pp 122–123.

[68] See Dencik, L., Hintz, A., and Cable, J. (2016). Towards Data Justice? The Ambiguity of Anti-surveillance Resistance in Political Activism. *Big Data & Society*, 3(2), 205395171667967. Schrock, A.R. (2016). Civic Hacking as Data Activism and Advocacy: A History from Publicity to Open Government Data. *New Media & Society*, 18(4), 581–599. Milan and Gutiérrez (2015).

[69] See Tactical Tech (nd). *Home*. https://tacticaltech.org

[70] Chenou, J.-M. and Cepeda-Másmela, C. (2019). #NiUnaMenos: Data Activism From the Global South. *Television & New Media*, 20(4), 396–411.

[71] D'Ignazio, C. and Klein, L.F. (2020). *Data Feminism*. The MIT Press.

[72] Auf der Suche nach Corona im Berliner Untergrund (2021) *Zerforschung*. https://zerforschung.org/posts/auf-der-suche-nach-corona-im-berliner-untergrund/

4

Representing Science

When I was a teenager in 1990s UK – a period which coincided with the height of public obsession with the TV show *Friends* – a series of adverts for hair products circulated featuring the *Friends* actor Jennifer Aniston. Famous for her glossy hair and 'Rachel cut', in the ads she talks about falling in love – with a shampoo. The most famous version of the advert doesn't only feature Aniston, however: she breaks off with the immortal words 'Here comes the science bit – concentrate!', allowing the ad to segue into an animation representing the shampoo in question's innovative technology and a (male) voiceover that explains this technology. According to the scriptwriters, this allowed them to include the 'obligatory scientific message' with humour and a sense of fun.*

I have forgotten a lot of things from this period of my life, but – for better or worse – this advert is not one of them. And perhaps it is ripe for re-analysis. There's a lot to reflect on in it: that the 'science bit' is framed as comprehensively separate from Aniston; that it is disembodied; that it involves jargon (the product contains Ceramide-R!); that there is the suggestion, through Aniston's winking 'concentrate!', that it is boring or at least demanding. The advert is just one example of the ways in which science and technology populate culture, not only through the technologies we use or how we imagine the role of science in society, but in popular media, consumer culture, and entertainment. Technoscience permeates leisure as well as politics.

This chapter explores some of the ways in which it does so, and how it comes to shape our shared visions and imaginations not just of science but also of collective life and the future. In it we look at how science is represented in news and entertainment media, and at some of the subtle ways – such as that shampoo advert – in which it forms part of

* It's not clear why it's obligatory – though of course that they say that already tells us something about advertising culture at the time, and the authority of science.[a]

public culture without us really being aware of it. At least some of these manifestations of technoscience may seem trivial: who really cares about what is represented in a shampoo advert, after all? Unlike some of the other topics we explore in the book (the politics of expert advice, disasters and crises, policy and governance), these informal or fictional versions of technoscience are easy to dismiss as unimportant. It's therefore essential to note that my starting point is that this is exactly not the case. It is vital that we pay attention to such presentations and representations of science, not just because they can work to shape our shared imaginations of what science is and the kinds of people who can work in it, but because in a very real way they constitute technoscience itself. If we cannot separate 'science' and 'society', then the versions of science that circulate in public, often almost unnoticeably, affect how science is practised.* We'll see examples of this throughout the chapter. We start, though, by looking at what we know about how science and scientists are represented in public media such as films, literature, and news.

Box 4.1: Stethoscopes as the 'hallmark of a doctor'

Popular images of technoscience have impacts not just on public perceptions, but on researchers and others involved in science and technology. Take, for instance, the way in which the stethoscope has become tied to public images of doctors. It is seen as a central signifier of being a doctor (as an image search for 'doctor' shows), even though newer technologies have replaced its use in many cases.[1] Having spent time observing medical students during their hospital training, Tom Rice describes the pride that they took in acquiring and wearing their stethoscopes, which to them captured 'the essence of being a medic'.[2] At the same time he found anxieties around the de-skilling that was said to be taking place as doctors relied more on tests and less on embodied skills such as using the stethoscope. Even though many medics seemed to wear it to signal their status as much as to use it in everyday encounters, '[w]ithin the discourse of the "death of the stethoscope", the stethoscope itself has come to represent, or stand for, endangered qualities and skills' – such as the cultivation of close relationship between a

---

* As one immediate example, there are many stories (some surely apocryphal) about how those working in science were inspired to go into it, or even to try and develop particular technologies, by science fiction. According to these accounts, if it weren't for Star Trek's transponders, we might never have had mobile phones.[b] (Though it should be noted that such accounts generally present a rather straightforward imagination of invention: an idea – or science fiction prototype – leads a particular person to develop a particular technology. We might question whether technologies actually develop in this way, or whether their creation is, perhaps, less linear than this model suggests – as discussed in Chapter 3.)[c]

doctor and their patient, and having a good 'bedside manner'.[3] They therefore represent much more than a simple tool to listen to the body.

How are science and scientists represented in public?

What do we know about how science is represented in public spaces and media, beyond our own experiences of shampoo adverts and the like? While such representations are often locally specific, there do seem to be some key themes that structure how science and scientists are imagined and portrayed, many of which date back to the early days of western science. Rosalind Haynes,[4] in a monumental study of representations of scientists in western literature from Chaucer to the end of the 20th century, identifies a number of stereotypes that recur throughout this work. One is the alchemist, or obsessive researcher – a figure that comes clouded in the mystique and secrecy of early alchemical practitioners, and which lies behind the notion of the 'mad scientist'. Another is the stupid virtuoso or absent-minded professor, another stereotype that dates from the early days of science and which plays on the image of the researcher who forgets everything apart from their science, including their familial or social responsibilities. Another again is the 'heroic adventurer'. This figure, different to some of the others, is portrayed more positively – though, as Haynes notes, in their guise as hero of space opera science fiction they tend to reproduce patterns of colonial invasion and appropriation.[5]

In all Haynes lists six of these stereotypes, most of which come with negative connotations. Her most important finding is perhaps not the stereotypes per se, but her argument that these images repeat over and over again in western literature and popular culture. They are, she suggests, 'archetypes that subsequently have acquired a cumulative, even mythic, importance'.[6] As such they are instantly recognisable, giving us a convenient heuristic for thinking about the nature of science and scientists. We know what to expect when we come across a 'nutty professor' in film or TV, or when we see a sinister, secretive researcher depicted in fiction, or when we watch a heroic lone scientist striding out into a new land (Jurassic Park, maybe). We know how the story ends.

Of course, the fact that such archetypes exist and circulate does not mean that they are uncritically accepted by ourselves or any other public audience. Rather, we might view them as resources. Producers of popular culture can draw on them in crafting stories (and may deliberately subvert them as they do so). Similarly, consumers* of popular culture might use them to

* It's worth noting here that throughout this chapter I use the idea of consumption as it has been described in cultural studies: something that is active, reflective, and conscious,

make sense of particular situations or narratives, whether in the context of fictional storytelling or elsewhere. However, few of us are likely to take such representations as a faithful depiction of scientific practice. Here there is a parallel with the so-called 'Draw a Scientist' test, where school children are asked to draw what they think a scientist looks like. They consistently draw images that look much like the archetypes Haynes identified – but, as science education scholar Joan Solomon has pointed out, what else would you expect from a relatively nonsensical question?[7] It is similar, she and colleagues write, to being asked to draw a dragon. To be aware of stereotypes does not mean that you uncritically accept them.

Box 4.2: The politics of medical illustration

What kinds of bodies are visible in public representations of technoscience? While scientists are a key category here, and tend to be represented in particular ways, it is also important to consider who else is depicted in images and other representations relating to technoscientific issues. A number of studies have shown that medical illustrations, for example, overwhelmingly depict bodies with light skin tones, a bias that 'reinforces whiteness as the assumed norm'.[8] Such taken-for-granted illustrations form a 'hidden curriculum' in medical education and training. Patricia Louie and Rima Wilkes write that in the context of depictions of race, 'while the formal [medical] curriculum emphasizes equality of care, this message is undermined when there is unequal representation in lectures, textbooks, case studies, and clinical training'.[9] It is for this reason that medical illustrators – such as medic Chidiebere Ibe – are seeking to create more diverse images of particular medical conditions or anatomy that can be used in medical training.[10]

Perhaps more pertinent is the question of how actual scientists tend to be represented in public, given that such representations are more likely to be understood as giving reliable insight into the nature and practice of research. Here there are long-standing concerns that the kinds of scientists who become publicly visible – and the ways in which they do so – do not well represent the culture of science as a whole. Analyses of media such as newspapers have shown that men are more likely to be quoted or featured in science reporting, and that women scientists tend to be framed in slightly different language, with an emphasis on aspects such as appearance, family status, and ability to combine their work with other interests.[11] The situation

rather than passive uptake of products or ideas.[d]

is extreme enough that there have been efforts to set up databases of women experts, so that journalists can find women sources and interviewees when covering science.[12] Similarly, projects such as 500 Queer Scientists seek to ensure greater visibility of LGBTQIA+ people in science, highlighting the stories and work of queer researchers in an effort to show the diversity of science in a way that is often not present in mainstream media.[13] Such examples emphasise the co-constitutive nature of these public representations and the technoscientific cultures they refer to. It matters if science is represented as primarily populated by White cis-heterosexual men – both for those within research, who may experience themselves as excluded from these versions of science, and for those thinking of careers in it.

There are thus efforts to encourage media reporting (in particular) that gives a more diverse picture of science. There may, however, be good reasons that some scientists (in particular those who are women or people of colour) may seek to avoid public visibility. Early studies of 20th-century public scientists – such as Margaret Mead or Carl Sagan – looked at how they rose to have a public profile, and at the effects on their careers.[14] While these visible scientists were more or less celebrated or controversial in public, the effects on their scientific trajectories were less positive, with their public-facing activities being seen as a 'pollution' or distraction. Today the situation in many countries is at least somewhat different. Celebrity scientists are often able to maintain both a public presence and successful scientific work, and are celebrated within science for doing so.[15] Indeed, the COVID-19 pandemic in many cases rendered this necessary: researchers at the cutting edge of vaccine development, virology, or epidemiology found themselves doing their research in the public eye. At the same time, however, it is clear that the benefits of public exposure are differential – not experienced equally by all researchers. On the one hand the ubiquity of social media, and the often virulent and polarised debate within it, has made the public sphere a sometimes deeply hostile environment for scientists. The rise of 'scientific popularism' discussed in Chapter 1 has connected scientific issues to political views and standpoints,[16] while the anonymity of social media has normalised extreme – even hateful – speech, with increased levels of abuse targeted at researchers with a public profile.[17] On the other hand, such hostility is more often directed at women, people of colour, and others who have not been traditionally considered natural inhabitants of the academy. As Tressie McMillan Cottom writes when reflecting on university encouragement of academics' public engagement, 'all press is good press for academic microcelebrities if their social locations conform to racist and sexist norms of who should be expert. For black women who do not conform to normative expectations of "expert," microcelebrity is potentially negative'.[18] While public communication and engagement is encouraged by many funders and research organisations, support for dealing with attacks on or through social media is often not in place. It is perhaps not surprising, then, that at least some scientists are increasingly hesitant about coming to public attention.

Box 4.3: Who is *not* represented in news media about science?

Just as important as the question of how scientists are represented in public is that of who doesn't become visible. There have been several egregious cases in which women of colour working in science have been cut from coverage of their own research, perhaps because they didn't fit the stereotype of what an 'authoritative' scientist looked like. One researcher, Rumman Chowdury, spent more than an hour explaining her work (and role as the founder of an artificial intelligence company) to a journalist, only to find that she was cut from the final article. As an article about the case notes, '[w]hen consulting people "on background", the general perception among journalists seems to be that these sources of background information do not need to be acknowledged, especially if there are no direct quotes included in the final piece'.[19] As such, 'women of color find their quotes on the chopping block, even when their ideas remain'. Part of the problem may stem from what one article called journalism's 'white supremacy problem':[20] in many contexts, journalism largely continues to be an elite profession lacking in diversity, one in which White journalists may have advantages over others in terms of getting access to jobs and stories. If journalism itself remains homogeneous, it is perhaps not surprising that the news it creates is similarly limited in the kinds of stories and voices that are covered. In the same way, academia itself has been criticised as structurally biased, giving White scholars advantages over others.[21] These dynamics will also impact who becomes visible in public media, in what ways, and with which kinds of reception.[22]

Media logics and representations of science

The rise of digital media and of particular cultures of interaction – in this case, those that normalise more extreme speech than we might use in face-to-face environments – is one example of the subtle ways that public representations of and engagement with science can be affected by the media in which they are present. Being concerned about the ways that science is represented in public thus means being interested not just in particular trends in content (such as gendered representations of scientists) but in the norms and logics of the venues in which these public representations take place. How do different forms of mediation shape how science becomes accessible to broader audiences?

Mass media such as newspaper reporting have received most attention with regard to this question. Historically, the formatting of news stories was extremely important: newspapers had a limited amount of space (so stories couldn't meander on indefinitely) and required complicated processes of page layout and typesetting (so were difficult to change quickly). These limitations have led to some entrenched conventions about what 'news' can and should look like – one instance being that it has an 'inverted pyramidal' structure, where there is a high degree of information density at the start of an article, and less important information (as decided by the journalist or their editor) further down.* Ideally, news writing should contain the 'who, what, where, why, and how' of its story in the first paragraph, and perhaps even in the first sentence – a feature that also caters to readers, who may well only scan the first part of a story. While the exact reasons for this convention are unclear, one advantage is that 'texts can be shortened from the end if the final layout requires it, and this saves time and staff during the editing process'.[23] News articles are therefore cut from the bottom. In the context of science stories, this can mean that material that journalists see as non-essential, but which scientists might see as important additional qualifying information, is either lost or remains at the end of an article, which readers may not reach.

Box 4.4: Is science reporting sensationalist?

Many scientists (and others) complain that news reporting of science stories exaggerates research findings, constantly claiming 'breakthroughs' or ignoring the limitations of particular studies. While this is certainly true in some cases, one large-scale study found that news reporting is actually more cautious about research findings compared to the original articles written by scientists.[24] Another study found that exaggerations in news stories tended to stem from press releases, rather than journalists – and press releases, as the article notes, are often written by university press officers in collaboration with the relevant researchers.[25]

There are also more intangible ways that news formats shape the presentation of science. Journalists write according to particular formatting requirements, but also in line with a sense of what is newsworthy. Given all the events that take place around the world every day, decisions must be made as to

* This, at least, is the theory. Though the idea of the inverted pyramid remains important in journalism and journalism studies, and is even taught to aspiring news writers, a quick glance at news stories will show that this is not a hard and fast rule. It also doesn't account for the different formats in which news now appears – from online newspapers to excerpts on social media – or the fact that science, in particular, is often presented not as news per se but as longer form articles which might emphasise the topic rather than its newsworthiness.[e]

what classes as 'news'. One way that journalists do this is by drawing on an internalised set of 'news values', which tell them what is newsworthy and what is less so. Such values help them with 'issue selection' (choosing what makes the news), and have been extensively described in journalism studies from the middle of the 20th century onwards.[26] They include geographical proximity (events that take place close are more newsworthy than those far away); influence (are celebrities or elites involved?); range or relevance (referring to how many people are affected by the story); controversy (the presence of conflict); and actuality (the 'new' aspect of news: has an event just happened?). These values (of which this is an abridged list: many others have been proposed) help shape what is reported on by particular venues, and what is not. It explains, for instance, why newspapers feature stories about research that comes from their particular national context; why a celebrity scientist making a statement about a scientific topic can turn it into a news story; and why the Higgs Boson is news only when its detection is a recent discovery. When a science story really blows up, and becomes headline news in its own right, it is because it meets one or more of these implicit criteria to a fault – perhaps by affecting almost everyone around the world (as with the COVID-19 pandemic).

News values highlight that news is made, not given. It is possible to imagine forms of media reporting about the world that follow other logics – something that is perhaps especially important in the context of science, where news values ensure that particular kinds of stories about science are told more than others.* The idea that newsworthiness should involve elites and new breakthroughs can lead to a focus on scientific superstars, for instance, even in cases where the reality of scientific practice is slow, painstaking, and collaborative. An emphasis on conflict (as well as the journalistic convention of 'balance' as a means of ensuring objective reporting) can mean that dissensus is at times manufactured or over-emphasised – something which might make for a 'better' news story, but which can poorly represent the science in question (a dynamic also discussed in Chapter 6, in the context of climate change). And the importance of relevance means that topics that closely touch our lives (such as medicine and health) tend to be better covered than other fields, especially those that seem to be arcane or abstract.

Just as importantly, journalists' and editors' ideas about what is newsworthy in relation to science mean that some science-related issues are rarely

* Indeed, we can see these demonstrated today through social media, where much more specialised forms of reporting and public discussion are possible (such as on blogs, or science-themed Facebook pages or TikTok channels). Different logics apply in this space with regard to what is worth reporting on – just one example of how digital media are changing the wider media landscape.

discussed in public venues. Felicity Mellor has coined the term 'non-news values' to refer to these cases, where 'editors and journalists ... invoke an implicit set of values about what science news does not look like'.[27] The examples she discusses include funding sources, connections to the military, and limitations and uncertainties: all of these are, she argues, systematically under-reported in press coverage of science, but represent important aspects of research that deserve public attention and scrutiny. Journalistic conventions that these things are not newsworthy mean, for instance, that how research is funded, and by whom, gets little attention, even when this might affect how we think about the findings; similarly, a downplaying of the limitations of particular studies means that scientists' claims are often not questioned. This is despite the fact that many science journalists see their role as being 'watchdogs', and as holding science publicly accountable.[28]

Non-news values (and news values) are just one example of the ways that the conventions and logics of public media can shape how technoscience is represented in them. These seemingly innocent norms, shaped by particular historical circumstances, come to configure which aspects of science are made visible, and how it can be thought about and discussed (as certain versus uncertain, for instance, or as carried out by particular researchers rather than by groups or communities). Most research in this area has focused on our oldest mass media – newspapers and journalism, in particular – but the ability of particular formats to structure content applies to other public spaces, too. Museum exhibitions, for instance, are often understood as presenting authoritative, 'finished' science, which can render it difficult to convey uncertainty or controversy.[29] As we have already seen, digital media – and in particular social media – are also normalising forms of communication that may be more aggressive than that which would take place in other contexts. In all of these spaces we find different structures, logics, and values being used to present and discuss knowledge than those that are used in science – something that can trigger complaints from scientists about science communication and public science, when the norms that come to define how science is represented in public differ from those used within scholarly communication.

Representing the future

The structures and formats through which technoscience are presented in public thus affect how it is represented, and in a very real way come to shape what it is. But it is not only scientists, particular technologies, or aspects of science that are constituted in this way. Other aspects of society are tangled up with public representations of technoscience, and are themselves constituted – made visible and thinkable – in particular ways. One key example of this is how we collectively think about the future. Potential futures and projections (of where technology is going, or how we might live in the coming years, for instance) may seem intangible, but are often represented in tandem with technoscience. They are thereby made concrete in specific ways through such representations.

Take an example from the early days of the COVID-19 pandemic, in the first half of 2020. This was a moment when scientific information exploded into the public sphere, and was mobilised by different actors to make recommendations or regulations. One chart that circulated in both social and mass media was titled 'Flattening the curve'. It depicted two different curves of predicted COVID-19 numbers against time: the first, in a scenario where no protective measures were taken, was sharper, with a peak that was higher and that arrived sooner; the second (in a scenario in which measures such as lockdowns or physical distancing were imposed) was gentler but lasted longer. The point was that the gentler curve did not exceed a horizontal line marked on the chart: that of healthcare system capacity. Both curves were prospective, based on modelling of different scenarios. The technical expertise behind such modelling was used to project what might happen, under particular circumstances, and thereby to (implicitly) make recommendations.

The chart is a good example of how public representations of science are entangled with other things, in this case with the question of different possible futures and how desirable these might be. Its message is clear: we don't want a future of health system chaos, and therefore we need the protective measures that can 'flatten the curve'. Indeed, potential futures are often tangled up with technoscience in public media. Consider how science fiction represents possible future worlds, or the promises that are made for emerging technology such as artificial intelligence, or the ways in which we are encouraged to plan for or seek to mitigate potential futures (in the context of climate change, for instance). Technoscience in public frequently has a prospective dimension.

Box 4.5: Planning for the deep future

Planning for the future is central to how contemporary societies function, and technoscience plays a central role in this. One space where this is particularly clear is that of energy, including the management of waste from nuclear reactors, which remains dangerous for thousands of years. How should this material be stored and handled when it will endure into futures that are astonishingly distant to our current societies? This question has been treated in a number of media, such as the 2010 documentary *Into Eternity*, which explores the construction of the Onkalo waste repository in Finland and the choices its designers must make.[30] It is clear that while technical expertise is central to these discussions, it does not produce clear-cut answers. Timothy J. Foley, who studied debates around the Onkalo repository, argues that these discussions reveal different imaginations at play: '[t]he progressive imaginary asserts that future technological advancement will mitigate natural threats, while the precautionary imaginary is skeptical that technology will reliably solve pressing future challenges'.[31] Technical advice therefore comes bundled together with certain assumptions, whether those are that technoscientific progress will solve future problems, or that we cannot rely on this and must plan accordingly.

A number of authors have suggested that a future orientation is central to the functioning of contemporary societies more generally – that collectively we use ideas about possible futures, more than our understanding of the past or present, to guide actions and choices. As John Urry writes, 'many hold the future to be a better guide to what to do in the present than what happened in the past'.[32] This may seem obvious until we start to interrogate what the future *is*. Is it ever knowable, and by what means? Scholars of time such as Barbara Adam have suggested that 'the' future should not be taken for granted: the notion can be understood in different ways. Much contemporary future-thinking frames it as an empty space, one that is 'waiting to be filled with our desire', but this ignores the ways in which the future is embedded in socioeconomic, political, and environmental processes.[33] Similarly, Vincanne Adams and colleagues have described the ways in which anticipating the future has become integral to daily life, bringing us* to live under a regime in which we are called to prepare for possible futures and to optimise ourselves for them.[34] Anticipation, they say, defines the early 21st century, illustrating this with the way in which prevention has become central to public and reproductive health (we safeguard our reproductive health through particular preparations, for instance, or seek to prevent disease through vaccination).

If representing and preparing for the future is central to contemporary societies, then technoscience is inevitably part of this. Science, technology, and innovation are associated with progress, to the extent that there is a deeply entrenched idea that research will lead to new technologies and thus to social benefits. As discussed in previous chapters, for large parts of the 20th century basic research was framed as the goose that laid golden eggs, and it

* Who is 'us' and 'we' in these contexts? In using this general language – society, we, us – I am mirroring the language of many of the authors I am quoting, who are concerned with diagnosing trends on the level of western society or even the world as a whole. But I can feel that I am a bit uneasy doing so. Even large-scale trends are experienced in very diverse ways.

was expected, as David Guston and Kenneth Keniston write, that funding such research would result in 'a steady stream of discoveries that can be translated into new products, medicines, or weapons'.[35] But technoscientific expertise also plays a role in identifying possible futures. 'Future studies' involves different kinds of scholarship, but includes work that seeks to map out potential future scenarios based on current trends. In processes such as scenario workshops (which are also discussed in Chapter 8), groups of stakeholders collectively discuss the processes that define the present, and attempt to extend these into different future situations. Importantly, in the case of scenario workshops the output of these discussions is not one but *multiple* potential futures. Different choices or developments may result in one future or the other; the future is not predefined, then, but is still open to being changed (even beyond the scenarios that are presented). Those participating in such workshops – often policy makers or others with decision-making power – are able to act to take the future in more desirable directions.*

Box 4.6: Designing convincing futures

The RAND Corporation – now a non-profit policy advice organisation – and Shell were two early pioneers in scenario planning. Initially based on techniques that emerged from Second World War military needs, such as systems analysis and quantitative modelling, the technique moved towards narrative accounts and vignettes of possible futures.[36] Importantly, such scenarios are framed as aiding discussion and reflection, rather than making predictions or promises. 'Scenarios are not about predicting the future', notes a RAND blogpost. 'They are, rather, a rigorous and methodical way to consider several imagined future situations, or contexts, which could come to pass.'[37] Based on assessment of societal drivers and trajectories, scenario-building seeks to present plausible futures, allowing policy makers (and others) to reflect on preferences around those possibilities and the values and desires that these signal.[38] While it has its roots in highly technical methods, scenario planning is now no longer solely an elite activity, but has been incorporated into public deliberation on technoscientific futures (as discussed in Chapter 8).

---

* Of course, the question of what is a 'desirable' future is not necessarily straightforward. This is why, as we will discuss in Chapter 8, there have been efforts to gather more diverse groups of participants for such scenario workshops. What is consensually seen as desirable by a group of policy makers may be different to the aspirations of groups from outside politics and the civil service.

While anticipating and planning for plausible futures has been one key area of scholarship in future studies, studying how the future is represented in public venues is another. Such work has been labelled the sociology of expectations, and explores how collective imaginations of different futures are created and circulate.[39] Again, these futures are generally tied to technoscience in the sense that they are (imagined as) enabled by new technologies, innovation, or knowledge. This work has distinguished between expectations, on the one hand, and visions, on the other, where visions are broader, representing not just specific future developments but whole future worlds. Expectations may take a number of different forms: they may circulate as prototypes (such as of self-driving cars), as graphics or charts (like the 'flattening the curve' COVID-19 chart discussed earlier, or visualisations of potential technologies like nanobots), or even as funding programmes. All of these render possible futures concrete, as well as implicitly suggesting whether they are positive or negative. Research funding programmes, for instance, often imply that particular technologies *should* be developed, because they *will* bring about desirable outcomes. (Consider strategic funding for areas such as synthetic biology, data technologies, or the European Commission's 'Grand Challenges' programme: all offer funding for particular areas of technoscience because benefits are imagined or promised through them – as in the statement that synthetic biology will 'transform how we grow food, what we eat, and where we source materials and medicines'.)[40] Expectations and visions are therefore not neutral, but make promises or suggest threats to be avoided.

Constituting society

Why does any of this matter? We are, after all, used to seeing representations of the future in everything from adverts to movies. Why should we be particularly attentive to how technoscience becomes tangled up with such representations? One answer is that expectations are *performative*. In representing the future they have very real effects on the present, from directing research funding in particular directions to becoming self-perpetuating, taken-for-granted promises about the future. Their presence can therefore close down discussion of other possible futures, or hinder reflection on whether this is a direction we want to take as a society. Another answer relates to hype, and what has been called the 'ethics of promising'.[41] It is usual for there to be hype and high, perhaps even overblown, expectations of new or emerging technologies – indeed, the 'hype cycle' is now an established phenomenon, one that technology developers may make use of to strategically boost their ideas at the right moment.[42] Such developers anticipate that after a period of hype there may be a 'trough of disillusionment' where the bubble bursts, before the technology moves towards more stable development and uptake (the 'plateau of productivity'). But this doesn't account for the ethical dimensions of such hype. What is at stake in the creation of expectations and promises about technoscience?

The first decades of the 21st century have given us some answers to this question. We have seen excessive promising about technological innovation – particularly in the context of the Silicon Valley tech industry – result in real harm. The now defunct business Theranos is perhaps the prime example. Its founder Elizabeth Holmes was able to garner huge financial and political support for the revolutionary 'lab on a chip' technology she was promising, despite her limited experience with such technologies and minimal proof of concept. The 'lab on a chip' would, the claim was, be able to run multiple tests for everything from cholesterol to indicators of diseases such as cancer from just a drop of blood, using a highly compact device. Despite the fact that the technology never worked as promised – in fact never came close to doing so – Theranos prospered and its blood tests were rolled out to a public market, meaning that real people, with real health issues, were affected by its shortcomings. While the company eventually came crashing down, and Holmes convicted for fraud, commentators have argued that its culture of 'fake it till you make it' is indicative of tech business more generally, not just this one instance of it.*

Similarly, self-driving cars – particularly those designed for the consumer market – are another example of a technology that is over-promised and subject to minimal regulation. As Jack Stilgoe has described, too many involved in the technology choose to downplay the challenges of such innovation, ignoring the complex ways that cars fit into transport infrastructures that are governed by social norms and values that are not easily transposable into software.[43] Those promoting self-driving cars often ignore these complexities, making promises of full automation that have repeatedly not been kept, and do so with very little accountability, even when there are fatalities. Such hype may also have less visible, but no less insidious, effects by directing funding and interest away from other less hyped, but potentially more effective, areas. Morgan Ames has argued that tech-driven projects such as 'One Laptop Per Child' suck up huge amounts of financial support by making promises of the benefits of technological innovations, but fail because they ignore the realities of the contexts in which they are seeking to intervene.[44] In the case of One Laptop Per Child, not only did the 'crank-powered' laptop never work as promised, but its vision of providing poor communities with laptops without support for their use and repair, engagement with local educators, or knowledge of children's actual situations

* There are now huge amounts of reporting on Theranos in different forms, including a book by John Carreyrou, the reporter who eventually broke the story on it. I have found the podcast *The Dropout*[f] and discussion of the case on *The Received Wisdom* especially helpful.[g]

meant that it had deeply paternalistic overtones. Hype often functions to reinforce, not solve, inequality – and it therefore deserves critical reflection on the part of those who produce it, and those of us who encounter it.

The whole point of hype is, of course, that it is flashy, and that we notice it. But technoscience is also present in public in subtle, perhaps even invisible, ways. We should also pay attention to how technoscience works to shape society without us noticing, as well to the promises that are being made about the futures that technoscience could enable. One example of this subtlety are the visualisations, graphics, or charts that circulate as taken-for-granted representations of reality, and which are therefore rarely opened up for critique or debate. As an example we might return to the 'flattening the curve' graph representing potential COVID-19 futures discussed earlier. We've already seen how this presents two different futures, and implies that one is more desirable than the other. We could consider, though, what else it is doing: it is presenting just two futures, no more, and therefore renders any other possibilities invisible. It frames the capacity of a healthcare system as the most important (in fact, the only) determinant of whether a possible future is desirable or not. And it suggests that COVID-19 infection numbers can be straightforwardly modelled, and that they will progress in predictable ways. It is therefore conveying a message about both the predictability of infection rates, and the power of scientific expertise to effectively model and understand it.

We might agree with these assumptions – but the point is that they are assumptions, views about the world that are being conveyed as the chart circulates but which are never explicitly stated or opened up to debate. Such charts therefore hold rhetorical power. They persuade us of certain things without foregrounding them. We thus start to get some sense of the ways in which technoscientific knowledge and products can structure our understanding of the world without us necessarily noticing, conveying not just information about particular issues or facts but whole imaginations of science, society, and our place in these.*

It is not only charts that function in this way. Visualisations also operate rhetorically to convey much more than we might immediately see. Martyn Pickersgill has studied how neuroscience is represented in public, suggesting that the ways in which brain scans are visualised are compelling even when we don't fully understand them, and that they can legitimise overly simplistic models of the brain (mapping complex conditions, characteristics, or

* Indeed, some Science and Technology Studies work would argue that this is less about conveying particular ideas about the world, and more about enacting the world in a particular way. Such knowledge products are therefore able to constitute, and not just describe, reality.[h]

processes onto particular areas, for instance).[45] More generally, numbers and statistics are another 'literary technology' (we met the term in Chapter 2) that allow scientific findings to travel and to persuade audiences of their validity. 'Quantification is a form of rhetoric that is especially effective for diffusing research findings to other laboratories, languages, countries and continents' writes Theodore Porter.[46] But '[t]his does not mean that mathematics is a neutral language'. Quantification and statistical analysis inevitably involve removing and compressing the complexity of the world, smoothing out and averaging away anything exceptional and telling a story of both tidy data and tidy realities. For example, another form of visualisation that has risen to prominence since the start of the COVID-19 pandemic is colour-coded maps representing vaccination or infection rates. Such maps travel well, and often feature colours that immediately signal the status of a situation: red areas are bad, while blues and greens suggest something more positive. But – when it is nation states that are the focus of this colour coding – they inevitably smooth over regional and local differences, or even those between household and household. In doing so they present nations as internally homogeneous but also as independent, clearly bordered entities – an idea that could be questioned at a time of globalisation and international work and travel. In addition, the data that lies behind such mapping efforts can be hard to identify and assess: they thus also occlude the difficulties and contingencies of developing reliable datasets around infection and vaccination.

Numbers, charts, and images – the background hum of technoscience in public – should therefore not be taken for granted. They present versions of the realities that they apparently represent, but these versions might always be questioned. As with any representation, they operate as persuasive devices, conveying background assumptions and ideas about the world. This is not negative per se: it would be impossible to create a representation that does not do this. But the subtlety with which this happens deserves our attention. We may agree with the version of the world that is reinforced by COVID-19 maps or charts – one in which the nation state is a key unit of comparison, scientific models are always accurate, and vaccination rates are more important than rationales for embracing or rejecting such vaccination (for instance). But it is worth noticing what such representations are doing, how they are shaping shared public imaginations of the world and of technoscience – and what they are *not* reporting, and therefore rendering invisible.

Box 4.7: Towards responsible quantification

Numbers are central to our lives, and to how our societies function. Statistics – on anything from numbers of hospital beds to numbers in higher education – are used to compare nations and measure progress. Such numbers have rhetorical power: in the case of UK government statements about COVID-19, for instance, round numbers – such as '100,000 tests per day' – were used to signal political achievements or milestones. Precise numbers – like 1,308,071 – were used to suggest precision, but also to 'hide' bad news.[47] Numbers are also central in less visible ways, for instance in the 'big data' that social media companies collect, sell, and use to render their products more sophisticated. This reliance on quantification has led some to call for a more ethical approach to it. Writing in the context of the COVID-19 pandemic and the use of mathematical models in policy making on this, Andrea Saltelli and colleagues argue that '[m]odellers must not be permitted to project more certainty than their models deserve; and politicians must not be allowed to offload accountability to models of their choosing'.[48] Their manifesto calls for more responsible approaches to modelling in order to better serve society. They suggest that there is a need for an ethics of quantification, not least because '[n]umbers capture our attention; they illuminate the part of reality that is being made numerical, and fatally push those parts into the background that come without the clothing of numbers'.[49] The use of numbers in policy or public discourse should not be allowed to hide from view aspects of the world that are not readily quantified, but which may be equally important.

Conclusion

This chapter has explored some of the ways in which technoscience (and those who work in it) are represented in public, how such representations are formed through particular conventions and logics, and the work that they do (for instance in solidifying certain futures over others, or hyping technologies or businesses). All of this is important, I have suggested, because public representations do not operate in isolation from the spaces, practices, and people they refer to, but function to co-constitute them. When media reporting of science is dominated by male scientists (for instance) this shapes imaginations of who does and might work within technoscience, while conventions such as news values mean that certain forms of science and technology become publicly visible over others. Technoscience is not the only thing affected by this: public representations of science and technology also help to constitute other aspects of society, from potential futures to nation states. The images of science and technology that circulate in popular culture and the media are therefore a powerful means through which technoscience and society come to (together) take on meaning and form within particular societies. Even shampoo adverts signal that science has particular characteristics, and relates to wider public culture in particular ways (that it is separate from it, and inaccessibly difficult, for instance).

What I haven't discussed thus far is how such representations are consumed, interpreted, and negotiated by their audiences. The next chapter focuses on such processes of reception and engagement. How do laypeople engage with and participate in technoscience?

References

[a] See their account of the campaign at GoodPilot (nd). *L'Oreal.* http://www.goodpilot.co.uk/loreal.html

[b] See Strauss, M. (2012, 15 March). Ten Inventions Inspired by Science Fiction. *Smithsonian Magazine.* https://www.smithsonianmag.com/science-nature/ten-inventions-inspired-by-science-fiction-128080674

[c] See, for instance, a now classic account of the development of the bicycle: Bijker, W.E. (1997). *Of Bicycles, Bakelites, and Bulbs: Toward a Theory of Sociotechnical Change.* MIT Press.

[d] See Du Gay, P., Hall, S., Janes, L., Mackay, H., and Negus, K. (1997). *Doing Cultural Studies: The Story of the Sony Walkman.* SAGE.

[e] See Pöttker, H. (2003). News and Its Communicative Quality: The Inverted Pyramid – When and Why Did It Appear? *Journalism Studies,* 4(4), 501–511.

[f] Jarvis, R. (Host), Dunn, T., and Thompson, V. (Producers) (2019). *The Dropout* [Audio podcast]. ABC News. https://abcaudio.com/podcasts/the-dropout

[g] Parthasarathy, S. and Stilgoe, J. (Hosts) (2019–present). *The Received Wisdom* [Audio podcast]. Shobitap.org. https://shobitap.org/the-received-wisdom/2022/1/11/episode-22-theranos-medical-devices-and-indigenous-knowledge-on-climate-change-ft-kyle-powys-whyte. See also Carreyrou, J. (2018). *Bad Blood: Secrets and Lies in a Silicon Valley Startup.* Pan Macmillan.

[h] See, for instance, Mol, A. (2002). *The Body Multiple: Ontology in Medical Practice.* Duke University Press.

[1] FitzGerald, E. (Producer) (2017, 6 August). The Stethoscope (No. 270) [Audio podcast episode]. *99% Invisible.* Radiotopia. https://99percentinvisible.org/episode/the-stethoscope

[2] Rice, T. (2010). 'The Hallmark of a Doctor': The Stethoscope and the Making of Medical Identity. *Journal of Material Culture,* 15(3), 287–301, at p 296.

[3] Rice (2010), pp 293, 288.

[4] Haynes, R.D. (1994). *From Faust to Strangelove: Representations of the Scientist in Western Literature.* Johns Hopkins University Press.

[5] See also Higgins, D.M. (2021). *Reverse Colonization: Science Fiction, Imperial Fantasy, and Alt-Victimhood.* University of Iowa Press.

[6] Haynes (1994), p 3.

[7] Solomon, J., Duveen, J., and Scott, L. (1994). Pupils' Images of Scientific Epistemology. *International Journal of Science Education*, 16(3), 361–373.

[8] Louie, P. and Wilkes, R. (2018). Representations of Race and Skin Tone in Medical Textbook Imagery. *Social Science & Medicine (1982)*, 202, 38–42, at p 39.

[9] Louie and Wilkes (2018), p 38.

[10] See Ibe, C. (nd). *Home*. https://www.chidiebereibe.com

[11] Chimba, M. and Kitzinger, J. (2010). Bimbo or Boffin? Women in Science: An Analysis of Media Representations and How Female Scientists Negotiate Cultural Contradictions. *Public Understanding of Science*, 19(5), 609–624. Mitchell, M. and McKinnon, M. (2019). 'Human' or 'Objective' Faces of Science? Gender Stereotypes and the Representation of Scientists in the Media. *Public Understanding of Science*, 28(2), 177–190.

[12] For instance, The Women's Room [@thewomensroomuk] (nd). *Posts* [X profile]. https://twitter.com/thewomensroomuk

[13] 500 Queer Scientists (nd). *Home*. https://500queerscientists.com

[14] Goodell, R. (1975). *The Visible Scientists*. Little, Brown & Company.

[15] Fahy, D. (2015). *The New Celebrity Scientists: Out of the Lab and into the Limelight*. Rowman & Littlefield.

[16] Mede, N.G. and Schäfer, M.S. (2020). Science-Related Populism: Conceptualizing Populist Demands toward Science. *Public Understanding of Science*, June, 096366252092425.

[17] O'Grady, C. (2022). 'Overwhelmed by Hate': COVID-19 Scientists Face an Avalanche of Abuse, Survey Shows. *Science*, 2022. https://www.science.org/content/article/overwhelmed-hate-covid-19-scientists-face-avalanche-abuse-survey-shows

[18] Cottom, T.M. (2015). 'Who Do You Think You Are?': When Marginality Meets Academic Microcelebrity. *Ada New Media*, 7. https://scholarsbank.uoregon.edu/xmlui/handle/1794/26359

[19] Djanegara, N.D.T. (2022). Why are Women of Color Erased from Their Own Science Stories?, *The Objective*. https://objectivejournalism.org/2022/01/why-are-women-of-color-erased-from-the-science-stories-they-helped-create/

[20] Willems, A. (2022, 27 January). Pipelines of Exclusion on the Climate Change Beat. *Study Hall*. https://studyhall.xyz/pipelines-of-exclusion-on-the-climate-change-beat/

[21] Bates, K.A. and Ng, E.S. (2021). Whiteness in Academia, Time to Listen, and Moving Beyond White Fragility. *Equality, Diversity and Inclusion: An International Journal*, 40(1), 1–7.

[22] Finlay, S.M., Raman, S., Rasekoala, E., Mignan, V., Dawson, E., Neeley, L., and Orthia, L.A. (2021). From the Margins to the Mainstream: Deconstructing Science Communication as a White,

Western Paradigm. *Journal of Science Communication*, 20(1), C02. https://doi.org/10.22323/2.20010302

[23] Pöttker (2003), p 510.

[24] Pei, J. and Jurgens, D. (2021). Measuring Sentence-level and Aspect-level (Un)certainty in Science Communications. *arXiv preprint arXiv: 2109.14776.*

[25] Sumner, P., Vivian-Griffiths, S., Boivin, J., Williams, A., Venetis, C.A., Davies, A., et al (2014). The Association between Exaggeration in Health Related Science News and Academic Press Releases: Retrospective Observational Study. *BMJ*, 349(dec09 7), g7015–g7015.

[26] Badenschier, F. and Wormer, H. (2012). Issue Selection in Science Journalism: Towards a Special Theory of News Values for Science News?, in Rödder, S., Franzen, M., and Weingart, P. (eds) *The Sciences' Media Connection: Public Communication and Its Repercussions*. Springer Netherlands, pp 59–85.

[27] Mellor, F. (2015). Non-News Values in Science Journalism, in Rappert, B. and Balmer, B. (eds) *Absence in Science, Security and Policy: From Research Agendas to Global Strategy*. Palgrave Macmillan UK, pp 93–113, at p 107.

[28] Davies, S.R., Franks, S., Roche, J., Schmidt, A.L., Wells, R., and Zollo, F. (2021). The Landscape of European Science Communication. *Journal of Science Communication*, 20(3), A01.

[29] Macdonald, S. and Silverstone, R. (1992). Science on Display: The Representation of Scientific Controversy in Museum Exhibitions. *Public Understanding of Science*, 1(1), 69–87.

[30] Madsen, M. (Director) (2011). *Into Eternity* [Film]. Magic Hour Films.

[31] Foley, T.J. (2021). Waiting for Waste: Nuclear Imagination and the Politics of Distant Futures in Finland. *Energy Research & Social Science*, 72, 101867, p 3.

[32] Urry, J. (2016). *What Is the Future?* Polity Press, p 2.

[33] Adam, B. and Groves, C. (2007). *Future Matters: Action, Knowledge, Ethics*. Brill.

[34] Adams, V., Murphy, M., and Clarke, A.E. (2009). Anticipation: Technoscience, Life, Affect, Temporality. *Subjectivity*, 28(1), 246–265.

[35] Guston, D.H. and Keniston, K. (1994). *The Fragile Contract: University Science and the Federal Government*. MIT Press, p 2.

[36] Turner, F. (2006). *From Counterculture to Cyberculture: Stewart Brand, the Whole Earth Network, and the Rise of Digital Utopianism*. University of Chicago Press.

[37] See Hughes, M. and Virdee, M. (2022, 28 January). Why Did Nobody See It Coming? How Scenarios Can Help Us Prepare for the Future in an Uncertain World. *RAND*. https://www.rand.org/pubs/commentary/2022/01/why-did-nobody-see-it-coming-how-scenarios-can-help.html

[38] Ramírez, R., and Selin, C. (2014). Plausibility and Probability in Scenario Planning. *Foresight*, 16(1), 54–74.

[39] Konrad, K., Van Lente, H., Groves, C., and Selin, C. (2017). Performing and Governing the Future in Science and Technology, in Felt, U., Fouché, R., Miller, C.A., and Smith-Doerr, L. *Handbook of Science and Technology Studies*. MIT Press, pp 465–493.

[40] Voigt, C.A. (2020). Synthetic Biology 2020–2030: Six Commercially-available Products That Are Changing Our World. *Nature Communications*, 11(1), 6379.

[41] See Fortun, M. (2005). For an Ethics of Promising, or: A Few Kind Words about James Watson. *New Genetics and Society*, 24, 157–174.

[42] Brown, N. (2003). Hope against Hype: Accountability in Biopasts, Presents and Futures. *Science Studies*, 16(2), 3–21. One version of the hype cycle has been monetised by Gartner Consulting; see their account of it here: Gartner (nd). *The Gartner Hype Cycle*. https://www.gartner.com/en/research/methodologies/gartner-hype-cycle

[43] Stilgoe, J. (2020). *Who's Driving Innovation? New Technologies and the Collaborative State*. Springer International Publishing.

[44] See Ames, M.G. (2019). *The Charisma Machine: The Life, Death, and Legacy of One Laptop per Child*. MIT Press. Toyama, K. (2015). *Geek Heresy: Rescuing Social Change from the Cult of Technology*. Hachette UK.

[45] See Pickersgill, M. (2013). The Social Life of the Brain: Neuroscience in Society. *Current Sociology*, 61(3), 322–340. Dumit, J. (2004). *Picturing Personhood: Brain Scans and Biomedical Identity*. Princeton University Press.

[46] Porter, T.M. (1992). Quantification and the Accounting Ideal in Science. *Social Studies of Science*, 22(4), 633–651, at p 644.

[47] Billig, M. (2021). Rhetorical Uses of Precise Numbers and Semi-magical Round Numbers in Political Discourse about COVID-19: Examples from the Government of the United Kingdom. *Discourse & Society*, 32(5), 542–558.

[48] Saltelli, A., Bammer, G., Bruno, I., Charters, E., Di Fiore, M., Didier, E., et al (2020). Five Ways to Ensure That Models Serve Society: A Manifesto. *Nature*, 582(7813), 482–484, at p 483.

[49] Saltelli, A., Andreoni, A., Drechsler, W., Ghosh, J., Kattel, R., Kvangraven, I.H., et al (2021). Why Ethics of Quantification is Needed Now. UCL Institute for Innovation and Public Purpose, Working Paper Series (IIPP WP 2021/05), p 1. https://www.ucl.ac.uk/bartlett/public-purpose/wp2021-05

5

Public Engagements

The previous chapter explored some of the ways in which technoscience is represented in public and popular culture, finding science in shampoo adverts and in promises about future technologies, and observing scientists both in their stereotypical guises and as real people who may be squeezed out of public media because of assumptions about what researchers look like. One thing that wasn't discussed was how such portrayals of science are received. How do publics engage with such representations, and with science and technology more broadly?

This question is the focus of this chapter, which covers how laypeople consume, engage with, protest, and otherwise negotiate technoscience. Of course, these aspects often overlap: even when we consume* forms of science communication such as science documentaries or books, for instance, we are also actively making sense of scientific knowledge, fitting it into our existing knowledges and understandings. Any communication process is active, even those that focus on the transfer of information. When I give lectures to students (or as I write this book), I might like to think that I am seamlessly transferring knowledge from my mind to those of others, but in reality they (and you) are taking in some aspects and not others, fitting ideas into pre-existing frames or concepts, disagreeing with or rejecting some things, and all in all making sense of the content I discuss in their (your) own ways. When we encounter non-scientists engaging with technoscience we should therefore expect sophisticated negotiations of its content rather than passive absorption. Indeed, as we saw in Chapters 2 and 3, we should also expect to see technoscience being made – co-constituted – in non-scientific spaces. This chapter therefore overlaps in substantive ways with

* As a reminder, I am drawing on the language of cultural studies in talking about consumption, framing it not as a passive, mindless process, but as something that is carried out actively and consciously.[a]

several others, highlighting themes of active public engagement with technoscience and the intersections of scientific and other knowledges that we see throughout.

An initial example demonstrates some of these dynamics. Health communication is one important way in which technoscientific knowledge becomes visible in our lives, perhaps especially during times of crisis (such as the COVID-19 pandemic) but also more generally, as we are exhorted to stop smoking or get vaccinated or are given advice about how to manage conditions or illnesses.[1] But changing health behaviours through communication is notoriously difficult to achieve effectively (it is generally not because people are unaware of the dangers of smoking that they continue to do so, for instance[2]). Scientific advice enters complex social contexts in which we may choose to ignore that advice or to prioritise other things (such as the pleasures of a cigarette). In addition, medical knowledge may intersect with lived experience or diverging value judgements in ways that lead to its rejection. In the environmental breast cancer movement, for example, those affected by breast cancer question a dominant biomedical approach that emphasises individual and lifestyle factors, arguing instead for the importance of acknowledging and researching the ways in which toxic environments – harmful chemicals in everyday products or in particular geographical sites – can cause cancer. Based on their experiences of toxicity, and on a concern that individualising risk and responsibility ignores the complicity of corporations in allowing toxins into the environment, activists 'transform personal experience into scientific knowledge and then into political action', as Stephen Zavestoski and colleagues write.[3] Medical knowledge and advice thus intersects with lived experiences (of cancer and its causes) and values (the rejection of 'an individualized approach … that lays blame on women, rather than the political and social structures that allow them to be exposed to carcinogenic chemicals').[4] Here, as in other healthcare contexts, scientific knowledge is made sense of (and at times rejected or protested) in light of personal experience, priorities, and values.

The rest of this chapter explores these dynamics of how non-scientists consume and otherwise participate in technoscience, beginning by asking: how are the public representations of science discussed in the previous chapter used by their audiences?

Consuming science

What do we know about how technoscience is consumed by public audiences? As already noted, one central theme in this context is that public engagement with technoscience is always active. None of us are empty vessels, or blank slates, to be 'filled up' with the scientific knowledge that we might encounter in popular culture or through science communication.* This not only relates to interpretation, but to how non-scientists seek out and engage (or not) with technoscience. A vital starting point for thinking about the reception of and engagement with technoscience in public is thus asking not what science and scientists gain from such encounters – the question of whether public audiences take the 'right' messages from representations of science – but what particular publics get from these representations. Why do people view, read, visit, or otherwise explore technoscience?

Unfortunately, we know less about this than we could do, in part because much research has focused on examining scientific literacy and other measures of knowledge. Research that looks at how and why laypeople consume technoscience is much thinner on the ground. The research that does exist paints a picture of the complex ways in which technoscience is folded into everyday life, becoming meaningful through other concerns or interests.[5] Engagement with science purely for its own sake seems rather rare: rather, individuals consume science because it is useful to them (for instance in supporting their professional development, or because it relates to personal health concerns), because it provides emotional pay-offs such as feeding curiosity or giving a sense of excitement, or because such engagement is somehow pragmatically useful (it's the school holidays, and going to a science festival is something you can do as a family). In one study, Sharon Macdonald looked at how visitors to a science museum moved around (and talked about) a particular exhibition, and at how they explained their visit.[6] '[T]his science exhibition at any rate', she writes, 'is not being visited primarily for the fact that it is "science"'; instead, visitors framed their trip as one of a number of possible family or tourist activities, 'something that you do' with your kids or when you're in London. Similarly, they made sense of the exhibits through discussions about 'grandmother's kitchen, the cat that was run over two years ago, which trousers no longer fit, recipes for jam' (for instance).[7] Scientific content was folded into, and made sense of through, the dynamics of everyday life.

Other studies have similarly demonstrated that technoscience is not encountered in isolation, but within particular contexts, and serves specific purposes in these. Studies of engagement with science on social media, for instance, indicate that such engagement may be used to build communities

* This has been a key theme in research on and practice of science communication for some time. I have written about this literature quite extensively elsewhere, so don't want to delve into it here – but in brief there has been a move from what is called 'deficit model' thinking, in which public audiences are understood as cognitively deficit, and where 'filling them up' with scientific information will result in positive attitudes to science, and more dialogic models of communication, where public knowledges are not only acknowledged within communication, but welcomed as being useful to science.[b]

oriented to particular identities (such as being a 'science lover'), thus allowing users to feel a sense of collectivity and belonging.[8] In the same way, Mike Michael writes that to 'know about black holes, chaos theory, cold fusion, xenotransplantation, the "ear-mouse" and so on is to perform a particular identity'; consuming science communication is, then, to engage with a particular form of popular culture so as to signal the kind of person that one is.[9] And ethnographic work suggests that much engagement with technoscientific knowledge is not framed as 'science' (which is often viewed as alien and intimidating) but is mundane and individual: an 'everyday muddling together of snippets from school science, breakfast-table science and controversial political science, [which] remakes it into something more creatively personal and lively'.[10] Such research suggests that we all make sense of science in our own ways, reaching for it when we need it, perhaps even ignoring it when doing so also serves us in some way (as we will see in Chapter 6).

Perhaps frustratingly, there is therefore rather little generic to say about public engagement with science, or about how laypeople form opinions about particular aspects of science. Such engagement is driven by personal interests and situations, and is contingent on these. We can certainly measure what people say about attitudes to or trust in science in surveys, but these largely tell us exactly that: how people respond to specific questions when asked in a survey research context.* In practice views and experiences of technoscience are mediated through culture and context. In working with minoritised communities in the UK, for example, Emily Dawson discusses the ways in which they experienced museum-based science communication as 'not for them': science communication was, she writes, 'configured around Whiteness and middle/upper class values'. In this context science and science communication were framed as pursuits for White, wealthy individuals (for instance through the stories told and voices heard within it), thereby excluding those with other backgrounds.[11] Similarly, attitudes to vaccination emerge in relation to a web of other factors, including prior experiences with medicine and the degree to which one trusts governments or other institutionalised forms of authority.† Publics engage with particular aspects of technoscience in the light of their histories, interests, and cultural and other contexts, and their relations and encounters with it will therefore always be specific rather than widely generalisable. (Indeed, this may well echo our own experiences. To think about how we personally encounter and engage with technoscience may be to remember particular medical

* If you are interested, the Eurobarometer offers one regular survey of public attitudes to science in Europe.[c] John Law also discusses the survey at length.[d]

† This is discussed further in Chapter 7.

encounters and the need to understand medical advice, informal leisure activities such as watching YouTube videos or going to science events, or an emergent political issue that led us to research the science behind it – for instance.)

Box 5.1: Non-scientists as science communicators

In an article in the *Journal of Science Communication*, Summer M. Finlay and colleagues argue that too little attention has been paid to the ways in which diverse publics themselves act as communicators of technoscientific knowledge.[12] In such cases non-scientists are not only adept at engaging with science, but find it important to share it with others, acting as mediators of scientific knowledge and reframing it for their communities. One example they give is of how Griots – 'West African troubadours, storytellers, historians, poets, praise singers and musicians, all rolled into one' – participated in public health communication during the Ebola crisis in the 2010s. This was vital, they write, not only because of the importance of music in West African culture, but because by communicating in their own languages the Griots were able to contextualise health information through their 'cultural legacies and inheritance of Indigenous knowledge'. They thus actively shaped communication about Ebola by combining it with their own knowledges, experiences, and cultures.[13]

Box 5.2: Encountering and resisting classification at borders

Borders are increasingly technologised – no longer only a matter of handing over a paper passport, but involving the collection of digital data. This is particularly the case for people on the move, such as those applying for asylum. In Europe, in particular, there is now a vast technological infrastructure designed to process people arriving at the border, compile information about them, and slot them into specific categories.[14] As Stephan Scheel writes, 'practices of border and migration control have increasingly become knowledge practices [involving] a growing number of identification and surveillance technologies'.[15] Digital data is central to these processes. In some places, for instance, data from asylum seekers' mobile phones is extracted and analysed in order to try and identify their country of origin. Importantly, this infrastructure is not encountered passively. Despite efforts to collect specific kinds of data, '[p]eople on the move toward Europe propose their own chains of actors, data, and metadata, as well'[16] – data that may be more meaningful to their own experiences or priorities. Similarly, in encountering efforts to extract information from phones, people on the move engage actively to understand and subvert technological processes such as AI-based dialect identification. 'They can mimic dialects when providing their sample for speech biometrics', notes Scheel. 'They can claim to not possess a phone. Or they can buy or borrow a relatively new phone that does not provide any clues about their country of origin and travel history.'[17] Engagement with border technologies is thus met by comprehension and agency of people on the move, and by insistence on their own meanings regarding their travel.

'Uninvited' participation in technoscience

Even in consuming the science that is present in popular culture – from museums to social media – non-scientists engage with it actively, making sense of it through existing frameworks and interests. This is even more the case when laypeople choose to participate in technoscience more directly. In contrast to the formal mechanisms for public participation in science outlined in Chapter 2 (such as science shops or constructive technology assessment) and discussed further in Chapter 8, there are a range of ways in which laypeople get involved in technoscientific knowledge production in what has been called 'uninvited' participation.[18] As with other forms of public involvement with science, unruly participation has a long history, and can take a number of forms. In this section I consider the (overlapping) categories of research 'in the wild', activism and protest, and the appropriation and use of scientific knowledge.

The notion of 'research in the wild' acknowledges that knowledge production, including that which self-defines as scientific knowledge production, can take place outside of formal institutions such as universities and research organisations. This means that laypeople carry out their own research, guided by their own interests and priorities, according to their own quality standards. One example might be participation in hackerspaces and DIY biology labs, where members of the public can access tools and equipment (including, in the case of DIY biology, molecular biology laboratory equipment) to carry out their own projects.[19] Here citizens can develop technological products (a number of start-ups have emerged from hackerspaces) or carry out research into, for instance, the DNA of sushi, how to reverse engineer expensive therapeutic drugs, or the identification of new antibiotics.[20] While involvement in hackerspaces is often about leisure and pleasure, for at least some hackers the appeal comes from doing science outside a mainstream institutional context, away from the pressure to publish, to follow particular kinds of career paths, or (perhaps) comply with institutional standards regarding safety or ethics. Beyond hackerspaces, however, we also find laypeople carrying out research that immediately affects them: on medical conditions, for example, or on cases of environmental harm in their localities. Indeed, the term 'research in the wild' first emerged from studies of patient groups who collaborate with scientists but who can be considered 'genuine researchers' in their own right.[21] Using photos, written accounts, and surveys (for example), patients (or those who care for them) document particular conditions, offering a parallel source of knowledge to that of institutionalised researchers. In some cases they work with such researchers: they are, Michel Callon and Vololona Rabeharisoa write, 'not merely content to produce useful and original knowledge of their own disease' but seek to 'establish contacts with specialists in order to work with them on an equal footing'.[22]

Box 5.3: The Flint water crisis as citizen science

Non-scientists may get involved in technoscientific research when they experience technoscientific problems that are not being dealt with by mainstream science. This may, for instance, be the case in the context of environmental pollution and harm, where citizens may notice problems that have not been picked up on (or that have been deliberately ignored) by the authorities. In 2010s Flint, a town in Michigan in the United States, residents began noticing changes in the quality of their water supply after it was switched to a different source. Officials from institutions such as the Michigan Department of Environmental Quality repeatedly dismissed their concerns, leading them to carry out their own research into water quality and to reach out to scientists to help them with this, activities which eventually led to widespread recognition of lead contamination and the declaration of a state of emergency. While the case is sometimes held up as an example of 'citizen science', the story is a complex one. As Benjamin Pauli writes in his account of the crisis, not only was the relationship between the activists and scientists often strained, for the activists the problems related as much to a deficit in democracy as to technical questions regarding water treatment. The case signalled a 'crisis of democracy. ... In Flint activist Claire McClinton's words, "If we control our water, it's not gonna get poisoned" '.[23]

Other research in the wild takes a more oppositional stance. While the term 'citizen science' can be understood in different ways – many of which simply involve laypeople in data collection or analysis and leave them little scope to actively shape research[24] – it may be used to refer to research that is carried out by citizens on their own terms, at times in rejection of or in opposition to mainstream findings. Gwen Ottinger has extensively documented the work of environmental activists who monitor air or water quality.[25] While such activists may engage with mainstream science – not least to contest its findings and to argue for stricter regulation – Ottinger argues that their research is carried out under slightly different epistemic – knowledge-producing – conditions. Activists are often concerned with rapid changes in environmental quality, for example, on the level of hours or days, while environmental scientists and regulators may average out such fluctuations. Similarly, regulators may demand knowledge that aligns with particular standardised practices, even when the use of these may be arbitrary or at least debatable. Ottinger therefore suggests that citizen scientists offer epistemic diversity on particular issues, and that their activities can point out the limitations of institutionalised science. '[O]rdinary people's participation in environmental monitoring', she writes, 'can help expand environmental regulators' knowledge about pollution'.[26] But this can only happen if such regulators are prepared to engage with, and learn from, the ways in which citizen scientists produce knowledge, and if they recognise the value of research in the wild in offering not just *additional* knowledge (citizen science is often framed as consisting of data collection or helping to 'fill in gaps') but a different kind of knowledge.

A similar point applies to scientific engagement with traditional or indigenous knowledge. While it is becoming increasingly common for ecologists and others to acknowledge the validity and utility of such knowledge, it is often framed as an addition to existing science, something that provides data but does not disrupt or challenge extant scientific frameworks and epistemologies. As Nicole Latulippe and Nicole Klenk write, though: 'Indigenous knowledge is not mere "data" that can be slotted into exogeneous western scientific models. As embodied practice embedded within a worldview, Indigenous knowledge is inseparable from the socio-cultural, political, legal and other grounded, largely place-based relations and obligations that give rise to holistic knowledge *systems*.'[27] As systems of knowledge, indigenous and traditional ways of thinking incorporate much more than a set of particular facts; rather, they often make explicit ideas about how and why knowledge should be utilised. Information does not stand in isolation, but is tied to (self-)governance and to the wellbeing of a community.[28] To do justice to knowledge from indigenous and traditional communities it is therefore vital to acknowledge their differences from mainstream science, as well as what they can add to it. As with the environmental activists that Ottinger describes, indigenous research and knowledge-making offer unique epistemic systems, and thereby resources for understanding the world.

Box 5.4: Indigenous knowledge and university ethical review

Ethical review of research projects is an essential part of scholarship in many research systems. In such processes, researchers must describe their plans, any ethical issues that they raise (for example, if they will be speaking with children or other vulnerable groups), and how they will respond to these. Ethical review is designed to guard against harm, exploitation, or abuse: principles such as informed consent seek to ensure that research participants are treated in a respectful and safe manner. But it is rooted in a very specific set of logics and a tradition of, in particular, biomedical sciences in western contexts.[29] Standardised ethical review has thus been criticised as Eurocentric and universalising, imposing general principles when in fact these don't exist. This may be a particular issue for indigenous scholars, whose work may operate within different paradigms to western science. Nlaka'pamux scholar Jennifer Grenz, for instance, writes that 'current standard requirements of ethics committees – such as providing the exact questions that we will ask Elders and knowledge keepers, and having fixed research objectives and methodologies – are not consistent with our ways of knowing'. Having to adhere to standardised protocols and imaginations of research is another form of colonisation; there is therefore a need, she argues, to cede control of the ethics of research to wider communities: '[t]he ethics of research projects between Indigenous researchers and Indigenous communities should be reviewed only by those communities'.[30]

Research in the wild – knowledge production by laypeople outside of institutionalised science – may overlap with activism: activists can produce counter-knowledges as part of their organising. But activism need not explicitly involve 'alternative epistemologies' such as those of the environmental activists described earlier. Technoscience may form the object of protests or activism, or become part of the activities of social movements, without the methods of science itself being challenged. Again, such activism has been particularly pronounced in the context of health, where people with particular conditions or illnesses have a clear motivation to seek healthcare that meets their needs, and to protest or organise when this is not accessible. For instance, early AIDS patients organised through the group ACT UP (AIDS Coalition to Unleash Power) to draw attention to a lack of research and progress in treating the disease, and to protest the way in which drug trials were being conducted.[31] ACT UP was a protest movement in the truest sense – it was noisy, disruptive, and used creative and theatrical approaches to raise awareness of AIDS* – but it also involved extensive engagement with institutional science and scientists. As Steven Epstein has described, a subset of activists gained

* It is instructive to compare ACT UP's activities with Iris Marion Young's discussion of activism, where 'the activist believes it is important to express outrage at continued injustice to motivate others to act'.[e]

'cultural competence' in the relevant research, and used this (among other strategies) to enter into the scientific conversations that were taking place: ultimately, they were able to bring about changes in research practice that better served the needs of HIV positive communities (such as less strict entry requirements for drug trials*). Since then, many other patient groups have used similar strategies to demand a 'seat at the table' with regard to medical research, and to raise awareness of particular conditions (including the breast cancer activists discussed at the start of the chapter). Gaining 'cultural competence', in particular, remains a key approach, in which patients or activists essentially *become* scientists in order to credibly engage with researchers, pharmaceutical companies, or policy makers – at times in ways that may introduce hierarchies or tensions with other activists.[32]

Box 5.5: What makes activism credible?

A central challenge for activists on technoscientific issues relates to the credibility of their claims and demands. On what grounds should they be heard? Monamie Bhadra Haines argues that credibility strategies should be understood as culturally specific rather than universal: what makes sense in one context might not in another. She describes the history of anti-nuclear activism in India, and the kinds of strategies and arguments that this has involved. While there are parallels with activism in western contexts – such as efforts to demonstrate technoscientific expertise and to mobilise this to support activist arguments – Bhadra Haines shows that what works in stable liberal democracies may not be effective in other political environments, and that Indian activists have adapted their approaches accordingly. Increasingly, they use 'guerilla' strategies that work with the state's bureaucratic structures, and with what it expects and allows from its citizens. 'If activists cannot hold the state accountable through science', she writes, 'they have attempted to anticipate what other kinds of arguments and modes of contention may gain traction'.[33] How activists can gain credibility is thus dependent on local structures and expectations regarding the relationship between technoscience, the state, and wider society.

* In practice it is debatable to what extent the ACT UP activism and engagement with mainstream science involved alternative epistemologies: the activists mobilised different value systems around how trials and treatments should be carried out, and managed to insert these into research practice, so we might say that they represented epistemic diversity. Given the importance of their becoming conversant with mainstream science, however, Epstein suggests that 'the capacity (or desire) of activists to pose genuinely epistemic challenges to biomedicine has been limited'.[f]

Box 5.6: Nothing about us without us

In 2021, a high profile UK study of genetics and autism spectrum disorder (ASD) was launched and almost immediately suspended. The project, known as the Spectrum 10K Study, had aimed to collect DNA samples and other health information from 10,000 individuals with some form of ASD. The group 'Boycott Spectrum 10K', which is led by autistic people, organised protests against it, arguing that the project had failed to consult those with ASD and that the benefits to the community were unclear. Critics not only had concerns about data protection and ethics, but argued that a much more in-depth discussion between the researchers and the ASD community was necessary to ensure that the research was in line with their needs. A *Nature* article about the controversy quoted one Boycott Spectrum 10K member as saying that 'it is not clear how the study will improve participants' well-being', and that its 'aim seems to be more about collecting DNA samples and data sharing'.[34] The case not only illustrates how research may be experienced as counter to particular values – such as the right of those with a condition to have a say in how it is studied – but an increasingly important assumption within research, particularly in studies relating to disability: nothing about us without us.[35]

Outside of health and medical research, activism and protest may also seek to resist or render controversial particular technoscientific developments. In Europe (in particular) there was a wave of protest regarding genetic modification of crops in the early 2000s, while during the COVID-19 pandemic scientific advice to socially distance, wear masks, or get vaccinated sparked public protests worldwide (as well as significant online debate and the circulation of mis- and disinformation[36]). Conversely, 'Marches for Science' during the 2010s sought to develop activism oriented to the protection and promotion of science. Climate activism has also involved public marches and acts of civil disobedience in an effort to draw attention to the science of climate change and to demand policy action on it. Importantly, research has shown that such protest cannot simply be categorised in terms of social movements being 'pro' or 'anti' science. Resistance to scientific developments is rarely about the science in and of itself, but rather about the ways in which it impinges on social and political identities, norms, or affiliations. Protest concerning genetic modification of agriculture, for instance, was often connected to concerns about how multinational corporations were coming to dominate food systems, while resistance to pandemic measures was connected to deep-rooted cultural ideals about the role of government in society, individual choice, or trust in institutions.[37] Similarly, pro-science activism such as the March for Science can be understood as a 'collective identity-building exercise' whereby groups who frame their identities around science – scientists and their supporters – could come together for 'solidarity and internal community building'.[38]

Activism and social movements connected to scientific issues can therefore be understood as emerging around wider questions than the science in and of itself: there is more at stake than any single technology, controversy, or aspect of research (indeed, it is often our very identities as citizens or private individuals that are at stake – what is threatened is *who we are* as societies). This takes us to a final way in which publics engage with science, through its appropriation and use for their own purposes (whether in activist or other contexts). As with other forms of uninvited participation, appropriation has a long history: historian Adrian Desmond describes how, in the 17th century, radical artisans engaged with and drew upon nascent evolutionary theory in order to bolster their political claims.[39] Such practices have continued, with a striking feature even of protests and rejections of mainstream science being a mobilisation of scientific language, findings, and approaches. Few activists present themselves as anti-science; instead, they 'collect data', 'conduct research', and 'consult publications' in a manner that, as researcher Anna Berg says of her engagement with anti-lockdown protesters during the COVID-19 pandemic, is strongly reminscent of 'discussions had in institutionalized academic contexts'.[40] Appearing scientific is therefore highly prized, including in contesting or challenging scientific arguments or issues, and scientific knowledge is a resource to be drawn upon in doing so.

This has been particularly well documented in the context of activist (and especially environmental) organisations, who frequently draw on science to justify and emphasise their claims and demands. While some of this knowledge may come from their own research (as in the case of patient-led 'research in the wild'), academic science is viewed as a key resource. Accounts of environmental organisations have shown that they assemble scientific information from a range of sources according to their needs: they may read reports and journal articles, talk to colleagues or scientists, or use in-house scientific resources (such as colleagues with a particular scientific background), but the aim is not to get an overview of a particular field or topic per se but rather to meet pragmatic needs within the organisation's activities.[41] Such use of science may simply demonstrate that they are credible ('science-based'), but it may also become part of their public communication activities as they explain and promote their positions. Just as we saw for laypeople's engagement with science more generally, and the consumption of science within leisure activities, the starting point is the organisation's interests and needs, not the science in and of itself.

Box 5.7: Engaging with science in and through activism

There is plentiful evidence that when non-scientists want or need to access technoscientific knowledge, they are adept at doing so. Non-experts can thus educate themselves to gain 'cultural competence' in scientific issues. One case of this is climate activism, and perhaps especially youth climate activism. Writing in 2020, Corrie Grosse and Brigid Mark note that '[t]here is a growing movement of youth across the globe engaged in school strikes for the climate, fossil fuel divestment, and direct action to protect their homes, land, livelihoods and cultures'.[42] Young people often learn about climate science in school, or are 'growing up with it as a daily part of life, due to the environmental changes and traumas they experience and/or the media coverage they see'.[43] At the same time it is clear that many of these social movements are informed by, and deeply engaged with, the science of the climate crisis. Collectives such as Fridays for Future organise around slogans such as 'unite behind the science', even framing themselves as communicators carrying out outreach to public audiences.[44] Activists thus engage with scientific literature in order to understand, and make demands relating to, the climate crisis.

Epistemic diversity and epistemic (in)justice

Surveying the different ways in which non-scientists consume and engage with technoscience has highlighted a couple of central ideas, the first of which I opened the chapter with: all of us as non-specialists encounter, use, and negotiate science and technology in active ways (rather than simply passively absorbing facts or adapting to technological developments). The other has been more implicit in my discussion so far and is worth drawing out further: this is the notion that there is not a single method or approach for creating robust, reliable, or useful knowledge, but many. Technoscientific approaches offer valuable insights into the world, according to their internal logics and methods, but, as we have seen, other knowledge systems also exist and have insights to contribute on particular issues and topics. In this chapter we have seen this demonstrated as I discussed indigenous knowledges and their relation to western science, described 'research in the wild' carried out by patients rather than professional researchers, and noted that there may be 'alternative epistemologies' within some forms of citizen science. In all of these cases knowledge production operates according to different logics, values, and assumptions to those embedded in mainstream technoscientific research, and in doing so meets slightly different needs or provides different kinds of insights.

We can therefore speak of epistemic diversity: there are different ways of making (reliable, useful) knowledge. As we will see further in the following chapter, this is the case within technoscience itself: different disciplines (or 'epistemic cultures') go about creating knowledge in different ways. Sabina Leonelli describes some of the differences that come to matter between such epistemic cultures, from underlying theoretical assumptions to the methods used, infrastructures relied on, and publication practices.[45] The central point in the context of this chapter is that reliable knowledge can be generated outside of mainstream technoscientific institutions and spaces, and that the systems used to develop it may differ in significant – and useful – ways to technoscientific approaches. For Gwen Ottinger, for instance, activist science can result in 'epistemic innovation, or the creation of new epistemic resources—concepts, categories, and metrics that help us understand the world and our experiences of it'.[46] Research that is done differently to institutionalised knowledge production can thus provide valuable resources both for such mainstream science, and for our understanding of the world more generally (as we have also seen in discussing the relation between indigenous knowledge and ecology in Chapter 2).

Epistemic diversity is connected to a related concept, that of epistemic justice (or injustice) – or, better, to a set of related concepts that circle around questions of justice with regard to diversity in knowledge systems. There has now been extensive discussion of ideas such as epistemic exclusion, epistemic oppression, epistemic (in)justice, and epistemicide.[47] While there are important differences in how these terms are framed and used, they point to a shared interest in 'forms of unfair treatment that relate to issues of knowledge, understanding, and participation in communicative practices'.[48] Such unfair treatment can take many different forms, from the silencing of particular forms of knowledge to a lack of access to useful information or the imposition of unfair hierarchies of knowledge systems, but there has been a particular emphasis on acknowledging the violent repression of knowledges from the Global South and other marginalised ways of knowing. In the context of the forms of public engagement with technoscience discussed in this chapter, the notion of epistemic justice encourages us to examine not just where non-scientists actively engage with technoscientific knowledge, but the extent to which their knowledges, approaches, and priorities may be being silenced, undervalued, or suppressed as they do so. In the context of environmental activism, for example, Gwen Ottinger describes how regulators show a 'lack of awareness or refusal to acknowledge the concepts marginalized groups themselves use to understand their experience', to the extent that there may be 'outright rejection of alternative ways of understanding how residents experience air quality'.[49] Mainstream technoscientific approaches may therefore refuse to acknowledge or engage with knowledge from other systems, even silencing or rejecting it. Similarly, in writing about epistemic injustice in global health research, Himani Bhakuni and Seye

Abimbola note that '[k]nowledge practices in academic global health typically privilege dominant groups, thus diverging from plurality and the need to defer to the local, internal, or emic knowledge and sensemaking of the individuals and groups whose systems and realities the field seeks to alter'.[50] In such cases injustice is done because the knowledge of particular (subaltern) individuals or groups is disregarded – something that is not only unfair, but simultaneously counterproductive given that, as Bhakuni and Abimbola write, it is the experiences of these groups that is the focus of the research.

Notions of epistemic diversity and (in)justice thus provide a language for engaging with the different forms of knowledge that come into focus around technoscientific issues, and sensitise us to how such knowledges are assessed and used (or not). They help us reflect on questions that we will encounter again in Chapters 6 and 9 (in particular): whose knowledge counts in specific situations, how is this contested or controversial, and how do these dynamics relate to wider structural inequalities?

Conclusion

This chapter has explored ways in which publics consume and otherwise engage with technoscience, emphasising that such consumption and engagement is always active, on the one hand, and that publics bring their own knowledges and epistemic practices to engagement with it, on the other. Thus we have seen that use of science communication (such as museums or science on social media) serves particular purposes, from nurturing a sense of identity to occupying one's family during the holidays, and that it is folded into everyday life and personal needs and interests. Similarly, non-scientists participate in technoscience in a variety of ways, from protest and activism to carrying out their own 'research in the wild'. While some such engagement may align with mainstream science, other instances involve different forms of knowledge and of epistemic practice, where there are diverging aims, standards, and methods to those of institutionalised research. The existence of epistemic diversity is thus a central point: not all knowledge-producing systems are the same, and forms of research in the wild such as activist data collection or indigenous knowledge should not be expected simply to function as an additive to institutionalised technoscience, something that 'slots in' to its paradigms and frameworks. Indeed, imagining that they should risks imposing epistemic injustice upon them.

The next chapter picks up on many of these ideas, both with regard to engagement (or disengagement) with science as relating to personal identity and to the idea of epistemic diversity. It considers how technoscience comes to matter in contexts of crisis and disaster, and how such situations are managed through technoscientific means. How can we think not only about different forms of knowledge, but non-knowledge, and what are the limitations of mainstream technoscientific approaches to this?

References

[a] See Du Gay, P., Hall, S., Janes, L., Mackay, H., and Negus, K. (1997). *Doing Cultural Studies: The Story of the Sony Walkman*. SAGE.

[b] See Davies, S. and Horst, M. (2016). *Science Communication: Culture, Identity and Citizenship*. Palgrave Macmillan.

[c] Eurobarometer (nd). *Latest Surveys & Publications. European Union*. https://europa.eu/eurobarometer/screen/home

[d] Law, J. (2009). Seeing Like a Survey. *Cultural Sociology*, 3(2), 239–256.

[e] Young, I.M. (2001). Activist Challenges to Deliberative Democracy. *Political Theory*, 29(5), 670–690, at p 673.

[f] See Epstein, S. (1995). The Construction of Lay Expertise: AIDS Activism and the Forging of Credibility in the Reform of Clinical Trials. *Science, Technology & Human Values*, 20(4), 408–437, at p 427.

[1] As indicative examples, a number of communication projects are described here: Tsanni, A. (2022, 15 February). African Scientists Engage with the Public to Tackle Local Challenges. *Nature*. https://www.nature.com/articles/d41586-022-00415-w

[2] Thirlway, F. (2016). Everyday Tactics in Local Moral Worlds: E-cigarette Practices in a Working-class Area of the UK. *Social Science & Medicine*, 170, 106–113.

[3] Zavestoski, S., McCormick, S., and Brown, P. (2004). Gender, Embodiment, and Disease: Environmental Breast Cancer Activists' Challenges to Science, the Biomedical Model, and Policy. *Science as Culture*, 13(4), 563–586, at p 572.

[4] Zavestoski et al (2004), p 564.

[5] My discussion here draws on a number of key texts: Falk, J.H. and Dierking, L.D. (2012). *The Museum Experience Revisited*. Left Coast Press. Jensen, E. and Buckley, N. (2014). Why People Attend Science Festivals: Interests, Motivations and Self-Reported Benefits of Public Engagement with Research. *Public Understanding of Science*, 23(5), 557–573. Fogg-Rogers, L., Bay, J.L., Burgess, H., and Purdy, S.C. (2015). 'Knowledge Is Power': A Mixed-Methods Study Exploring Adult Audience Preferences for Engagement and Learning Formats Over 3 Years of a Health Science Festival. *Science Communication*, 37(4), 419–451.

[6] Macdonald, S. (1995). Consuming Science: Public Knowledge and the Dispersed Politics of Reception among Museum Visitors. *Media, Culture and Society*, 17(1), 13–29.

[7] Macdonald (1995), p 19.

[8] Marsh, O.M. (2018). Emotional and Descriptive Meaning-Making in Online Non-Professional Discussions about Science. PhD Thesis. University College London.

[9] Michael, M. (1998). Between Citizen and Consumer: Multiplying the Meanings of the 'Public Understanding of Science'. *Public Understanding of Science*, 7(4), 313–327, at p 318.

[10] Solomon, J. (2013). *Science of the People: Understanding and Using Science in Everyday Contexts*. Routledge, p xiii.

[11] Dawson, E. (2018). Reimagining Publics and (Non)Participation: Exploring Exclusion from Science Communication through the Experiences of Low-income, Minority Ethnic Groups. *Public Understanding of Science*, 0963662517750072.

[12] Finlay, S.M., Raman, S., Rasekoala, E., Mignan, V., Dawson, E., Neeley, L., and Orthia, L.A. (2021). From the Margins to the Mainstream: Deconstructing Science Communication as a White, Western Paradigm. *Journal of Science Communication*, 20(1), C02. https://doi.org/10.22323/2.20010302

[13] See also: Deffor, S. (2019). Ebola and the Reimagining of Health Communication in Liberia, in Tangwa, G.B., Abayomi, A., Ujewe, S.J., and Munung, N.S. (eds) *Socio-cultural Dimensions of Emerging Infectious Diseases in Africa: An Indigenous Response to Deadly Epidemics*. Springer International Publishing, pp 109–121. Rasekoala, E. (ed) (2023). *Race and Sociocultural Inclusion in Science Communication: Innovation, Decolonisation, and Transformation*. Bristol University Press.

[14] Pelizza, A. (2020). Processing Alterity, Enacting Europe: Migrant Registration and Identification as Co-construction of Individuals and Polities. *Science, Technology, & Human Values*, 45(2), 262–288.

[15] Scheel, S. (2024). Epistemic Domination by Data Extraction: Questioning the Use of Biometrics and Mobile Phone Data Analysis in Asylum Procedures. *Journal of Ethnic and Migration Studies*, 1–20, at pp 4–5. https://doi.org/10.1080/1369183X.2024.2307782

[16] Pelizza (2020), p 280.

[17] Scheel (2024), p 14.

[18] Welsh, I. and Wynne, B. (2013). Science, Scientism and Imaginaries of Publics in the UK: Passive Objects, Incipient Threats. *Science as Culture*, 22(4), 540–566.

[19] See Davies, S.R. (2017). *Hackerspaces: Making the Maker Movement*. Polity Press.

[20] Talbot, M. (2020, 18 May). The Rogue Experimenters. *The New Yorker*. https://www.newyorker.com/magazine/2020/05/25/the-rogue-experimenters. Waag Futurelab (nd). DIY Antibiotics. *Waag*. https://waag.org/en/project/diy-antibiotics

[21] Callon, M. and Rabeharisoa, V. (2003). Research 'in the Wild' and the Shaping of New Social Identities. *Technology in Society*, 25(2), 193–204.

[22] Callon and Rabeharisoa (2003), p 198.

[23] Pauli, B.J. (2019). *Flint Fights Back: Environmental Justice and Democracy in the Flint Water Crisis*. MIT Press, pp 14, 20.

[24] Strasser, B.J., Baudry, J., Mahr, D., Sanchez, G., and Tancoigne, E. (2018). Rethinking Science and Public Participation. *Science and Technology Studies*, 32(2), 52–76.

[25] Ottinger, G. (2010). Buckets of Resistance: Standards and the Effectiveness of Citizen Science. *Science, Technology & Human Values*, 35(2), 244–270. Ottinger, G. (2022a). Misunderstanding Citizen Science: Hermeneutic Ignorance in U.S. Environmental Regulation. *Science as Culture*, 31(4), 504–529. Ottinger, G. (2022b). Responsible Epistemic Innovation: How Combatting Epistemic Injustice Advances Responsible Innovation (and Vice Versa). *Journal of Responsible Innovation*, 10(1), 1–19.

[26] Ottinger (2022b), p 2. See also Pauli (2019).

[27] Emphasis in original. Latulippe, N. and Klenk, N. (2020). Making Room and Moving Over: Knowledge Co-production, Indigenous Knowledge Sovereignty and the Politics of Global Environmental Change Decision-making. *Current Opinion in Environmental Sustainability*, 42, 7–14, at p 7.

[28] Whyte, K. (2018). What Do Indigenous Knowledges Do for Indigenous Peoples? In Shilling, D. and Nelson, M.K. (eds) *Traditional Ecological Knowledge: Learning from Indigenous Practices for Environmental Sustainability*. Cambridge University Press, pp 57–82.

[29] Wahlberg, A., Rehmann-Sutter, C., Sleeboom-Faulkner, M., Lu, G., Döring, O., Cong, Y., et al (2013). From Global Bioethics to Ethical Governance of Biomedical Research Collaborations. *Social Science & Medicine*, 98, 293–300.

[30] Grenz, J. (2023). University Ethics Boards Are Not Ready for Indigenous Scholars. *Nature*, 616(7956), 221. https://doi.org/10.1038/d41586-023-00974-6

[31] While ACT UP was an international movement, it has been most extensively documented in the United States: see one discussion here https://www.lrb.co.uk/the-paper/v43/n15/adam-mars-jones/good-activist-bad-activist, and the work of Steven Epstein: Epstein, S. (1996). *Impure Science: AIDS, Activism and the Politics of Knowledge*. University of California Press.

[32] See Epstein's (1996) discussion of ACT UP for one example of this.

[33] Haines, M.B. (2019). Contested Credibility Economies of Nuclear Power in India. *Social Studies of Science*, 49(1), 29–51, at p 29.

[34] Sanderson, K. (2021). High-profile Autism Genetics Project Paused Amid Backlash. *Nature*, 598(7879), 17–18, at p 17.

[35] The history of this term is discussed in: Harpur, P. and Stein, M.A. (2017). The Convention on the Rights of Persons With Disabilities as a Global Tipping Point for the Participation of Persons With Disabilities, in Leithner, A. and Libby, K. (eds) *Oxford Research Encyclopedia of Politics*. Oxford University Press, pp 1–22.

[36] Prasad, A. (2022). Anti-science Misinformation and Conspiracies: COVID–19, Post-truth, and Science & Technology Studies (STS). *Science, Technology and Society*, 27(1), 88–112.

[37] Silva, E.O., Dick, B., and Flynn, M.B. (2023). The Evil Corporation Master Frame: The Cases of Vaccines and Genetic Modification. *Public Understanding of Science*, 32(3), 340–356. Paul, K.T., Zimmermann, B.M., Corsico, P., Fiske, A., Geiger, S., Johnson, S., et al (2022). Anticipating Hopes, Fears and Expectations towards COVID-19 Vaccines: A Qualitative Interview Study in Seven European Countries. *SSM – Qualitative Research in Health*, 2, 100035.

[38] Riesch, H., Vrikki, P., Stephens, N., Lewis, J., and Martin, O. (2021). 'A Moment of Science, Please': Activism, Community, and Humor at the March for Science. *Bulletin of Science, Technology & Society*, 41(2–3), 46–57.

[39] Desmond, A. (1987). Artisan Resistance and Evolution in Britain, 1819–1848. *Osiris (Second Series)*, 3, 77–110.

[40] Berg, A. (2023). Anti-COVID = Anti-science? How Protesters against COVID-19 Measures Appropriate Science to Navigate the Information Environment. *New Media & Society*. https://doi.org/10.1177/14614448231189262

[41] Fähnrich, B. (2018). Digging Deeper? Muddling Through? How Environmental Activists Make Sense and Use of Science: An Exploratory Study. *Journal of Science Communication*, 17(3). Unander, T.E. and Sørensen, K.H. (2020). Rhizomic Learning: How Environmental Non-governmental Organizations (ENGOs) Acquire and Assemble Knowledge. *Social Studies of Science*, 50(5), 821–833.

[42] Grosse, C. and Mark, B. (2020). A Colonized COP: Indigenous Exclusion and Youth Climate Justice Activism at the United Nations Climate Change Negotiations. *Journal of Human Rights and the Environment*, 11, 146–170, at p 146.

[43] Grosse and Mark (2020), p 152.

[44] Rödder, S. and Pavenstädt, C.N. (2023). 'Unite behind the Science!' Climate Movements' Use of Scientific Evidence in Narratives on Socio-ecological Futures. *Science and Public Policy*, 50(1), 30–41.

[45] Leonelli, S. (2022). Open Science and Epistemic Diversity: Friends or Foes? *Philosophy of Science*, 89(5), 991–1001.

[46] Ottinger (2022b), p 2.

[47] Dotson, K. (2011). Tracking Epistemic Violence, Tracking Practices of Silencing. *Hypatia*, 26(2), 236–257.

Dotson, K. (2014). Conceptualizing Epistemic Oppression. *Social Epistemology*, 28(2), 115–138. Santos, B.S. (2015). *Epistemologies of the South: Justice Against Epistemicide*. Routledge. Settles, I.H., Jones, M.K., Buchanan, N.T., and Dotson, K. (2021). Epistemic Exclusion: Scholar(ly)

Devaluation That Marginalizes Faculty of Color. *Journal of Diversity in Higher Education*, 14(4), 493–507.

[48] Kidd, I.J., Medina, J., and Pohlhaus, G. (eds) (2017). *The Routledge Handbook of Epistemic Injustice*. Routledge.

[49] Ottinger, G. (2022). Misunderstanding Citizen Science: Hermeneutic Ignorance in U.S. Environmental Regulation. *Science as Culture*, 31(4), 504–529.

[50] Bhakuni, H. and Abimbola, S. (2021). Epistemic Injustice in Academic Global Health. *The Lancet Global Health*, 9(10), e1465–e1470, at p e1466. https://doi.org/10.1016/S2214-109X(21)00301-6

6

Knowledge in Crisis

> Reports that say that something hasn't happened are always interesting to me, because as we know, there are known knowns; there are things we know we know. We also know there are known unknowns; that is to say we know there are some things we do not know. But there are also unknown unknowns – the ones we don't know we don't know.
>
> Donald Rumsfeld, then US Secretary of Defence, in 2002[1]

Rumsfeld's comments – which came in the middle of a news briefing regarding the possible presence of weapons of mass destruction in Iraq – were largely treated with derision at the time, even winning a 'Foot in Mouth' award from the UK Plain English Campaign for the 'most baffling comment by a public figure'. The phrasing is, perhaps, tortuous, and the structure confusing. But (remarkably enough) the basic idea that Rumsfeld is trying to convey is an important one, and one that we will explore throughout this chapter. Knowing and not knowing are generally taken to be straightforward, binary categories: we know or do not know a particular fact. But in practice these categories have texture and nuance. As Rumsfeld says, there are different ways of not-knowing, and, as we will see in Chapter 7 in particular, knowledge itself can be fragile and contestable. In this chapter we explore some of this fragility, looking at what happens to technoscientific knowledge in times of disaster or crisis, as well as the ways in which both knowledge and non-knowledge are constructed through the intermingling of scientific, social, and political processes. We therefore examine the kinds of unknowing that Rumsfeld describes. How do we come to know some things, know that we don't know others, and are entirely ignorant of the existence of others again?

Knowing and not knowing

In the decades since Rumsfeld made his comments the field of ignorance studies has emerged. Its basic premise is that ignorance is not simply emptiness or lack, but a rich social space that emerges in particular ways and has particular uses. In scientific research, for instance, we know particular things and not others because of funding and scholarly priorities and interests, all of which operate to focus research on specific areas (we continue to be ignorant of, for instance, many aspects of women's health, because standard scientific models are generally male, or of diseases that dominate the South rather than the rich North[2]). David Hess writes about 'undone science' as a means of capturing not just the fact that there will always be lacunae in research, but that where these fall is often prompted by power dynamics and the priorities of elites. Those who might advocate for 'a broad public interest' (such as social movements or activists), he writes, often 'find that the research that would support their views, or at least illuminate the epistemic claims that they wish to evaluate, is simply not there'.[3,*] The things of which we are ignorant – from particular environmental issues to numbers of femicides[4] – are therefore the result of particular choices made by and in research and data collection.

As a field, ignorance studies takes for granted that the aspects of the world that we are ignorant of are worthy of study, and that it is important to examine how and why ignorance is present. But it has also explored the texture of ignorance using the typology that Rumsfeld started to hint at. Kristian Nielsen and Mads Sørensen suggest that we can identify four different kinds of ignorance: known knowns ('well established facts and evidence'); known unknowns ('things we know that we do not yet know'); unknown knowns ('things that we do not know we know … things that we for some reason do not want to know'); and unknown unknowns ('absolute non-knowledge in the sense that we do not even know that we do not have the knowledge').[5] The last category is perhaps of particular interest with regard to the ways in which technoscience becomes visible within society. As Nielsen and Sørensen point out, unknown unknowns are central to the ways in which deleterious 'unanticipated effects' may emerge from new technologies. The drug thalidomide was used by pregnant women in the 1950s because nobody imagined it could have effects on their unborn children; chlorofluorocarbons (CFCs) were used in aerosols for many years because assessing their effects on the atmosphere had not been considered relevant. Such cases render visible not just a lack of

* This might be compared with the alternative epistemologies mobilised by activists discussed in the previous chapter.[a]

knowledge regarding particular impacts of technoscientific developments, but a lack of frameworks with which even to imagine such effects. Risk assessment – something which we will discuss later in the chapter – goes some way towards engaging with potential impacts, but can by its nature only handle possibilities that are imaginable within existing technical frameworks (known unknowns, in the categorisation discussed earlier). The very real possibility of unknown unknowns that result in harm or undesired effects has been one impetus towards calls for broader means of assessing technologies, for instance by including a wider range of citizens and stakeholders as a way of extending imaginations concerning what might not be known.[6] The argument has been that those outside of science may be better at considering the unexpected, the things that could emerge that are entirely outside of existing frameworks of knowledge.

Box 6.1: Modelling creates both knowledge and ignorance

Computer models are central to many fields, from climate science to public health. In the case of climate research, models build on historical and contemporary data to simulate climate systems, make future projections, and explore potential impacts of climate change. They are central to both research and policy, but, as Stephen Bocking points out, they conceal as much as they illuminate. No model can include all relevant information, and all require their inputs to be structured in specific ways – as quantitative data, for example. Writing in 2004, Bocking argued that climate models had been developed in a manner oriented to gradual change, and that this had impacts on policy: '[t]aking their lead from models, discussions of scenarios and policy options tend to focus on gradual change, and do not consider events deemed to be unlikely'. Fundamentally, 'only those aspects of reality [that models] can represent are likely to be viewed as possible' in discussions around whatever is being modelled.[7] This means that any model must be contextualised, and interpreted with caution. They create knowledge about a particular phenomenon, but simultaneously ignorance, in that they render certain 'aspects of reality' invisible.

The notion that there are different approaches to ignorance, and ways of engaging with it, is central not just to arguments for the value of broader citizen participation in assessing technologies (which are further discussed in Chapter 8) but to understanding how different disciplines and scientific cultures themselves engage with the unknown. Just as scientific method is, in practice, not universal, but involves diverse 'epistemic cultures' that create knowledge in slightly different ways,[8] such cultures also integrate different approaches to ignorance. Stefan Böschen and colleagues describe three key epistemic cultures of ignorance within scientific practice: 'control-oriented' (such as in molecular biology), where the emphasis is on tightly delimiting as many aspects of research as possible and eliminating 'disruptive factors'; 'complexity-oriented' (such as in ecology), where there is an openness to the unexpected; and 'single case-based experience' (such as medical case histories), where a lack of completeness – and thereby a degree of ongoing ignorance – is expected.[9] This becomes significant not just with regard to interdisciplinary collaboration, where researchers may expect research partners from different fields to operate with the same basic approach to non-knowledge, but to the ways in which research enters public spaces and is taken up in scientific controversies. Control-oriented approaches are often dominant, bringing with them expectations of laboratory conditions and the control of variables; other kinds of knowledge – such as case studies or complexity-oriented approaches – may be framed as invalid when contrasted to them. In practice, however, Böschen and colleagues argue that a combination of epistemic approaches is necessary, alongside the recognition that control-oriented cultures are not well equipped to engage with the possibility of unknown unknowns in their objects of study.*

Why does any of this matter? Those who write about technoscientific ignorance are clear that there is a politics to how ignorance emerges and is discussed. We have already hinted at the ways in which the things of which we are ignorant are tied to the kinds of questions that we, as societies, choose to prioritise: 'undone science' includes women's health issues such as endometriosis and the environmental impacts of commonly used chemicals.[10] Similarly, engaging with different types of ignorance can point us to the ways in which it can become politicised, with only certain types of knowing or unknowing being considered grounds for action (such that case studies may be dismissed as anecdote, for example). While many of these debates seem to take place in good faith, with those involved unaware of their own ignorance (so to speak), the politics of unknowing also includes less honest engagement with ignorance. In particular there is evidence that ignorance can be strategic, serving particular purposes that, for instance, protect businesses from accountability or action, or that maintain the status quo. Daniel Lee Kleinman and Sainath Suryanarayanan have documented chemical company Bayer's negotiation of a controversy in which its products were implicated in a massive spike in honey bee deaths in the United States. Importantly – and unlike other cases of willed ignorance on the part of businesses – they note that there is no evidence of misconduct or malfeasance (they contrast the case with the story told in the film *Erin Brockovich*, where a company

* Again, there are connections here to the ideas about epistemic diversity covered in Chapter 5, and to questions of epistemic justice discussed in Chapter 9.

deliberately seeks to cover up a case of environmental contamination). But by 'playing by the rules' of standards for evidence within the US regulatory system, and thereby disregarding the accounts of beekeepers, Bayer was able to systematically dismiss any possibility that the situation required further investigation. The company was able to mobilise its substantial resources, Kleinman and Suryanarayanan write, in order to argue that 'the observational and "anecdotal" data produced by beekeepers is not official, recognized or legitimate knowledge', and thereby to protect itself from regulation. In essence, Bayer denies the validity of all accounts outside of a very specific epistemic structure, ultimately bolstering ignorance about honey bee die-off rather than working to resolve it.[11]

Box 6.2: Institutional ignorance in Himalayan hydropower

What are the impacts of ignorance? Amelie Huber explores this question in the context of hydropower development in the Himalayas, looking at how experts, government spokespeople, and citizens frame controversies around the construction and management of dam projects in northeast India. Uncertainty and unknowability are, she argues, used as a means of turning a 'blind eye' to the risks that hydropower involves, with the result that elites are able to benefit from its economic impacts. Here the choice to minimise or remain ignorant of the risks of intervening into ecologically complex environments allows residents' concerns to be dismissed. '[I]nstitutionalized ignorance about certain hydropower risks and impacts' has, she writes, 'enabled the unchecked construction of hazardous hydropower projects'. Even if this ignorance is not deliberate, it is certainly strategic: the 'hydropower community has used ignorance as an excuse not to act'.[12]

Even more egregious are the actions of industry in the context of climate science. Here and in similar cases, such as knowledge of the health impacts of smoking in the tobacco industry, there is evidence of manufactured ignorance, where businesses seek to maintain ignorance or to sow confusion.[13] Analysing publications from the fossil fuel corporation ExxonMobil, Geoffrey Supran and Naomi Oreskes show that even when internal documents acknowledged the reality of human-induced climate change, public-facing material sought to cast doubt on this and to emphasise uncertainty and non-knowledge. Similarly, ExxonMobil and other companies funded 'groups and individuals and participated in organizations that cast doubt in public on climate science'.[14] Other scholars have argued that, in seeking to foster a sense of public uncertainty regarding the science of climate change, such companies were able to strategically exploit media norms and logics (such as the news values discussed in Chapter 4), and in particular the notion of 'balance'. News media traditionally seek objectivity, achieving this by attempting to tell 'both sides of a story' (for instance with regard to a political controversy). Their efforts to give balanced accounts have, however, been utilised by the fossil fuel industry as a means of implying that there is a considerable body of evidence both for and against human-induced climate change, even when the scientific consensus was firmly in favour of the reality of the climate crisis. Even when journalists thought they were acting in the public interest – for instance by seeking a diversity of views on climate science issues, or by acting as a 'watchdog' on scientific claims – their actions have fostered ignorance and confusion by giving prominence to niche science, or by presenting climate research as a partisan, political issue.[15]

Ignorance can therefore be produced for particular self-interested purposes. This is not only true of industry: in everyday life, remaining ignorant of particular aspects of the world can also be expedient. What Charles W. Mills has called 'White ignorance' is one example of this, involving the systematic forgetting or ignoring of the continuing presence of White supremacy, colonialism, and racial injustice in order to maintain racial domination.[16] A lack of acknowledgement of colonial histories – a kind of a willed ignorance – continues to shape much European culture,[17] even while such histories continue to shape technoscience and its ignorances in diverse sites around the world. Numerous scholars have now written about the ways in which coloniality – both as history and as continuing practice – affects the kind of technoscientific development that can take place, and the ways in which this is realised. Writing about efforts to promote the transition to sustainable development around the world, for instance, Saurabh Arora and Andy Stirling note that 'racialised patterns of discrimination and appropriation are generally overlooked or sidelined, and often inadvertently condoned through normalised reproduction of white privilege'.[18] Ignorance of colonialism, racism, and White supremacy is thus systematically embedded into mainstream western cultures, including technoscientific practice, with the result that such structures are less visible to interrogation, challenge, or resistance. Similarly, Noemi Tousignant describes how colonial legacies, research into toxicity, and a lack of capacity to know about risk and harm come together in Senegalese toxicology. Despite a constant struggle for resources that would allow for more knowledge about pollution (and the ongoing comparison with better funded science in the Global North), scientists have 'improvised and imagined a more capacious and protective toxicology, thereby "refusing" the forms of not-knowing that threatened the "civicness" of their practice'.[19] Here toxicological ignorance is creatively resisted with the aim of preventing harm, with both the risks at stake and the science to understand and monitor them structured by Senegal's colonial history.

While 'White ignorance' and related forms of systematic forgetting of colonial violence are particularly insidious, it is important to note that ignorance is not always negative. Those in ignorance studies are clear both that a degree of ignorance is inevitable (in research and elsewhere), and that it can serve valuable social purposes.[20] Consider the category of unknown knowns, used to refer to things that we do not know that we know (such as tacit or taken-for-granted knowledge), as well as knowledge that is taboo or that we do not wish to know. Ignorance may here be used to construct a sense of identity or to maintain social bonds or relations of trust. In discussing how laypeople reflect on their ignorance of science, Mike Michael points to a high degree of sophistication in handling this: respondents know they are ignorant regarding some things, and are able to reflect on how and why that might be. Just as knowledge of science can be tied to identity and the kind of person one is (as we saw in Chapter 5), ignorance may be used in the same way; here, however, science is 'bracketed as other' and identity tied to the fact that one *isn't* the kind of person who knows about it.[21] Relatedly, ignorance may relate to a division of labour, and to the sense that it is the role of others to know about it. 'Knowing too much' might, in this case, be framed as impinging on their social role, or as implying that one didn't trust those knowers. Or, again, Michael reports that interviewees talked about simply not wanting to know some things – exactly what would happen in the event of a meltdown at a nearby nuclear reactor, for instance.[22] In the same way, participants in one study about health choices were often content to remain ignorant until consulting someone that they saw as an expert, such as a doctor, who would give them reliable information. 'Ignorance is bliss sometimes!', said one participant.[23] Indeed, there have now been extended discussions of the 'right to ignorance' in the context of genetic testing and the possibility of predicting the onset of disease.[24] In the context of conditions with no effective treatments, or where there may be stigmatisation or penalisation (such as higher insurance premiums), it is perhaps not surprising that ignorance is understood as preferable to knowledge.

Box 6.3: Ignorance and online mis- and disinformation

Ignorance has entered a new phase with the rise of digital and social media and, in particular, what has been characterised as an era of mis- and disinformation. Misinformation is the spread of false information, whether knowingly or unknowingly, while disinformation involves the deliberate choice to mislead or promote false information.[25] The internet contains vast quantities of both (as well as of reliable information). Some studies have indicated that social media lend themselves particularly well to the rapid spread of false information because, on the one hand, in online contexts 'people often share misinformation because their attention is focused on factors other than accuracy' (such as humour or partisan commitments)[26] and, on the other, false information endures in online spaces (rather than having the transient lifetime of a print newspaper article, for instance).[27] However, concerns about misinformation have also been criticised as an exaggerated 'moral panic', with research indicating that in practice social media users engage with misinformation online rather sceptically. To quote the title of one article, '[p]eople believe misinformation is a threat because they assume others are gullible' – in other words, users are personally sceptical when they encounter false information, but assume that other people are more credulous.[28]

Risk as non-knowing

Ignorance is thus everywhere. It is intrinsic to science, in which every line of research involves choosing to *not* know about other aspects of the world, is mobilised strategically to protect particular interests or to maintain the status quo, and may be a valuable means of maintaining social bonds or a sense of identity. The ubiquity of non-knowledge (and its close cousin, uncertainty) means that strategies have emerged to deal with it in scientific and public contexts. One of these is the notion of risk, which has become central to the imaginations and governance of many contemporary societies.

Risk is a shifting concept, difficult to define. Its everyday usage tends to refer to the possibility of threat or harm, but it has also stabilised in technical discourse as a means of quantifying and calculating potential forms of damage.[29] At the same time risk can be understood as something positive: 'risky' sports or behaviours can be a source of pleasure, while in academic contexts 'high risk high gain' research projects are understood as innovative studies that have the potential for significant advances in knowledge. In many ways ideas of risk have populated our lives and what it means to live in modern societies. Ulrich Beck, for instance, popularised the notion of the 'risk society', in which new forms of harm have emerged as a result of technoscientific progress, and where the distribution of risk has become central to dynamics of privilege and equity. 'In advanced modernity', he writes: 'the social production of *wealth* is systematically accompanied by the social production of *risks*. Accordingly, the problems and conflicts relating to distribution in a society of scarcity overlap with the problems and conflicts that arise from the production, definition and distribution of techno-scientifically produced risks'.[30]

In other words, industrial modernity has had unintended impacts: as well as resulting in social progress and technological development, the last centuries have seen environmental degradation, new forms of technically enabled exploitation, and the collapse of biodiversity (for instance). These forms of harm, these risks, are intrinsic to contemporary forms of wealth production, and their distribution becomes central to social organisation. For Beck, tension and inequalities previously confined to the distribution of resources – who has access to wealth and to the means to live well – have now partly shifted to the distribution of risk. Who gets to live safely, and who is at risk of harm from, for example, environmental toxins, extreme weather events such as flooding or fires, or radioactive waste? While Beck writes about the risks of late modernity in part being so insidious because they may be invisible or spread across long periods, with radioactivity being his paradigmatic example, at the same time he is clear that risk is not equally distributed. Wealthy countries are able to outsource their risks (in the form of dangerous by-products of industrial processes for example) to other countries, while the poorest in any country tend to be the ones most at risk of harm.

Box 6.4: The Agbogbloshie e-waste scrap market, Accra, Ghana

One central aspect of the 'risk society' is the exporting of risk and harm to lower income regions and countries. Waste disposal is a central instance of this, with vast sites for waste frequently located in the Global South, and waste of all kinds shipped from the North to be processed, recycled, or simply dumped. The Agbogbloshie dump in Accra, Ghana – 'one of the world's largest destinations for used electronic goods' – is one example.[31] Hundreds of thousands of tonnes of electronic waste reaches Agbogbloshie each year, and it has become 'a major scrap market that now also serves as the hub for e-waste processing and provides employment for 4000–6000 people'. Such processing has resulted in toxicity and environmental harm – one report called it one of the most polluted places on earth – and has led to calls for better oversight. At the same time, Grace Akese and Peter Little insist on the complexity of the situation. On the one hand, for most workers at Agbogbloshie 'immediate livelihood needs, housing needs, and the right to the city are more crucial than the potential long-term health risks posed by harmful e-waste practices'; on the other, 'localized historical processes ... make certain populations and places such as Agbogbloshie particularly susceptible to socioecological harm from e-waste processing'.[32] It is thus vital to explore the ways in which the 'exporting' of risk and harm (in this case in the form of dangerous waste) always takes a form specific to particular sites and their histories.

For Beck, risk therefore becomes a central logic structuring contemporary global politics. Though his thesis has been criticised – not least for focusing on a one-dimensional view of the nature and reception of 'risk'[33] – it is certainly true that risk and its assessment remain a key language for engaging with (emerging) technologies and their (potential) impacts. Public and policy discussion of genetic modification, nanotechnology, synthetic biology, or artificial intelligence (AI) have all tended to be framed around assessments of their 'risks and benefits', concomitant with an understanding that these should be satisfactorily balanced in order to ensure public acceptance. Where there have been public hesitations or concerns about such technologies, this has often been understood as relating to their risks – their potential for specific forms of harm – rather than to other aspects of their development and use. This is, perhaps, the central limitation of risk as a means of engaging with technoscientific ignorance: it is a technical notion that leaves little scope for engagement with concerns outside of the possibility of (physical) damage, and which therefore flattens many of the complexities and nuances of non-knowing discussed in the previous section. Brian Wynne has argued this particularly strongly, suggesting that critiques of the institutions that develop and govern science, and of the directions that research takes, have too often been diverted into 'argument over the reality or scale (or precise causes) of diverse threats to our security from technology's consequences, leaving aside the greater issues of what the proper human meanings, conditions, limits, and purposes of scientific and technological innovation should be'.[34]

Risk, Wynne suggests, focuses attention on 'the reality or scale' of potential harms, while public concerns or protest engage with a far more diverse set of anxieties, for instance regarding the 'conditions, limits, and purposes' of technoscientific development: what it is for, whether we need it, what innovation priorities should be, who gets to control technology, who benefits from it, and so on. To use risk as a singular frame for discussing such development therefore results in a democratic (and epistemic) deficit, in which the varied meanings that science and technology take on are rendered invisible, and critique of its purposes and directions – not just its effects – is neutered.*

Box 6.5: Should we be concerned about artificial intelligence's 'existential risks'?

Advances in AI technologies – and particularly the development of 'generative' AI that can generate original text, images, or other kinds of content – have been greeted both with enthusiasm and concern. Their actual and potential risks have been an important aspect of these discussions, with some in the tech industry warning of the 'existential

* This is, of course, a dynamic that we have seen repeatedly throughout this book. Public concerns are often reactions not to particular technoscientific developments in and of themselves, but to the ways in which they are framed, governed, and driven. See, for instance, discussion of vaccination politics in Chapter 7 and of activism around genetic modification of crops in Chapter 5.

risks' the technology might entail. Such risks are suggested to threaten humanity's survival as a whole, for instance in the case of a superintelligent AI deciding that it would be better if humans didn't exist. While some high profile scientists and industry figures have foregrounded such concerns, this focus has been criticised as a distraction from the very real harm that AI technologies are causing in the present.[35] 'End-of-days hype surrounds many AI firms, but their technology already enables myriad harms, including routine discrimination in housing, criminal justice and health care', write Alex Hanna and Emily Bender.[36] Similarly, Meredith Whittaker points out that many of the people warning about 'existential risk' are the people who actually have the capacity to act, if they really wished to:

> These are the people who could actually pause it if they wanted to. They could unplug the data centres. They could whistleblow. These are some of the most powerful people when it comes to having the levers to actually change this, so it's a bit like the president issuing a statement saying somebody needs to issue an executive order.[37]

Discussions of long-term hypothetical risks may thus direct reflection away from other, less dramatic but more urgent, forms of harm.

While risk is useful for managing some aspects of non-knowledge, then, it is not sufficient for engaging with technoscientific ignorance as a whole. (Indeed, the move to ignorance studies, and to increased discussion of 'unknown unknowns', is at times framed as a reaction to the prior dominance of risk as a frame for public discussion and assessment of technoscience.[38]) What is perhaps more productive is to explore the ways in which risk – as a flexible category of meaning – is mobilised in particular contexts. I have already mentioned that 'risky' sports can be understood positively; similarly, it is clear that in other contexts risk is understood as something to be welcomed, or that is simply 'part of the human project', as Deborah Lupton and John Tulloch put it.[39] Extensive research now indicates that most people do not engage with risk in the manner imagined by theorists such as Beck, as rational actors who carry out cost–benefit analyses based on potential for harm. Rather, risk is approached personally and flexibly, with multiple factors being brought into consideration when risk judgements are made: how much in control of a risky situation am I? Do I trust the experts (or others) who are warning of risks? How normalised is this risk in my community? Similarly, risk judgements are often emotional in nature. Risk researcher Paul Slovic makes a distinction between risk as analysis and risk as feeling, and suggests that certain kinds of risks are particularly emotionally resonant. In particular, those that provoke affects of 'dread' – low probability, high consequence events such as terrorist attacks or aeroplane crashes – are avoided more than others, even where the likelihood of harm might suggest that other behaviours (such as smoking) are more dangerous.[40] Technical assessments of risk may therefore differ from lay judgements in terms of what are 'risky' behaviours. In addition, 'risk perceptions are dynamic, changing … over the course of [individuals'] lifetimes or even from day to day'.[41]

Knowledge and non-knowledge in crisis and disaster

One reason that risk became such a dominant framework in the late 20th century was a sense that it was increasing and changing, to emerge in new, disturbing ways. Beck's work was prompted in part by anxieties about emergent risks such as radioactivity from nuclear waste or environmental degradation caused by the use of new chemicals. For many, the 20th century saw a decisive break in how humanity related to the world: rather than being to a large extent at its mercy, constantly threatened by extreme weather or the failure of crops, now the natural world was itself at threat from humanity's technologies and tools. Writing in the 1970s, philosopher Hans Jonas expressed this sense of a sea-change particularly clearly: previously, he writes, humanity's inroads into nature had been 'essentially superficial', but now:

> All this has decisively changed. Modern technology has introduced actions of such novel scale, objects, and consequences that the framework of former ethics can no longer contain them. … Take, for instance, as the first major change in the inherited picture, the critical vulnerability of nature to man's technological intervention – unsuspected before it began to show itself in damage already done.[42]

As the extract suggests, Jonas is concerned with the ethics of this transition, and how to move from a situation in which nature could (essentially) take care of itself to one where it needs to be considered an object of responsibility and care. But more generally these changes created a moment in which, relatively suddenly, the power of human-made technology seemed immense, capable of destroying both the world we collectively live in and humans themselves. After the techno-optimism of the early 20th century – a time of wonder chemicals and miracle cures – public reflection across the world increasingly came to focus on the harms that were emerging from technoscientific development.

Such disillusionment was heightened by a series of high profile disasters that, in stark contrast to being 'natural', were clearly linked to the development and use of technoscience. The use of the atomic bomb on Hiroshima and Nagasaki in 1945 and the chilling accounts and images that emerged of the after-effects of this was one turning point that highlighted the previously almost inconceivable power that science now held. The second half of the 20th century then saw, for example, the failure of the 'green revolution' (the use of pesticides and other techniques to increase crop yields that also caused a collapse in biodiversity), release of radioactive gases from the US Three Mile Island nuclear plant in 1979, toxic gas leakages at Union Carbide India Limited pesticide plant in Bhopal, India, in 1983, and an explosion and radiation leak at the Chernobyl nuclear plant in Ukraine (then part of the Soviet Union) in 1986. Such cases caused the deaths of thousands and debilitating conditions in hundreds of thousands, drawing attention both to the 'double-edged' nature of technology and to the industrial and organisational conditions under which it was realised. One analysis of the *Challenger* disaster (in which, in 1986, a US space shuttle disintegrated upon launch), for instance, indicated that it was caused less by a single faulty component and more by the culture of management at the National Aeronautics and Space Administration (NASA), in which, Diane Vaughan writes, the organisational response to a nascent technical problem 'was characterized by poor communication, inadequate information handling, faulty technical decision making, and failure to comply with regulations instituted to assure safety'.[43] Such cases made it clear that disasters were caused not only by one-off failures, but by the systems that technologies were embedded within.*

Indeed, this became a central aspect of imaginations of the nature of technological modernity and of disasters themselves. In the 1980s Charles Perrow introduced the idea of 'normal accidents': accidents, he argued, should now be considered a usual part of technological development and use, something to be expected as complex technologies (from ships to dams to nuclear power stations) are utilised. This is in part to do with the 'unknown unknowns' discussed earlier, and the ways in which consequences can never be fully anticipated. But Perrow also emphasised the systemic aspects of technologies – the ways in which they involve multiple components that interact with each other, and are embedded in similarly complex sets of interactions and components. This level of complexity renders it impossible to predict how failures in any one component will interact with other aspects of the system: 'No one dreamed', he writes (hypothetically),

> that when X failed, Y would also be out of order and the two failures would interact so as to both start a fire and silence the fire alarm. … Next time they will put in an extra alarm system and a fire suppressor,

* The same point has been made repeatedly regarding the Chernobyl and Bhopal disasters, in which poor maintenance, lack of funding, and efforts to cover-up the extent of the disasters all contributed to their effects.

> but who knows, that might just allow three more unexpected interactions among inevitable failures.[44]

Disasters can therefore never be fully predicted or prepared for, while simultaneously being inherent to complex technological systems.

Such high profile disasters, and reflection on them, have led to a whole field of scholarship around disaster management – a field that has in itself become a space of technical, and sometimes technocratic, expertise. While this research predominantly frames disasters as one-off events that cause substantial damage and that are focused on specific moments and places, the work I have discussed so far suggests a rather broader definition, one that emphasises the systemic aspects of disasters and their extended timeframes and geographies.[45] Such scholarship insists on the necessity of understanding disasters in context, and particularly on engaging with questions of justice and equity with regard to who is most affected by them. Work carried out by the Superstorm Research Lab (a mutual aid collective initiated after Superstorm Sandy hit New York in 2012), for example, identified narratives of 'two Sandys': one account that focused on temporary infrastructural damage and disruption of New York's status as a financial centre, and another that instead framed Sandy through 'ongoing, unequal social and economic conditions in place before the storm and exacerbated afterwards'.[46] The second narrative, which primarily emerged from grassroots organisations, highlighted the ways in which the effects of this particular weather event were in fact not only due to a particular moment of disaster, but to long-standing issues of poverty and neglect in particular areas of the city. Similarly, these impacts stretched forward, into the future, with recovery efforts complicated by pre-existing gaps in infrastructure or resources. As Max Liboiron notes, the second narrative suggests that 'alleviating wealth inequality and keeping public housing in good repair would be a step toward increasing resilience', rather than more complex preparations for potential future weather-related disasters.[47] In the same way other disaster researchers have argued for the need to better account for queer experiences of disaster, and to engage with the intersections of disaster events with diverse forms of marginalisation.[48]

These discussions emphasise the complexity of how disasters are experienced, and the ways in which even 'natural' disasters are entangled with pre-existing (technological) infrastructure. Even where crises are not directly triggered by failures of technoscientific systems, expert knowledge is central to how they are predicted, managed, and tracked. Aside from disaster specialists (whose work may involve advising on resilience or on recovery efforts), scientific data is often key to how disasters are made publicly meaningful – consider, for example, colour-coded maps of COVID-19 cases during the 2020 pandemic, or charts of radiation levels after the 2011 Fukushima nuclear disaster. Increasingly, this raises questions regarding whose knowledge counts in a crisis, and whether technoscientific expertise is well equipped to be the primary sensemaking tool for understanding a disaster. Many disasters now involve citizens intervening to create their own datasets or bodies of knowledge in opposition to those produced by scientific experts. After Fukushima, for example, hackers and citizen scientists created kits for homemade Geiger counters in order to measure radiation levels, producing data at a level of granularity not provided by official sources.[49] Similarly, those involved in the Superstorm Research Lab chart how grassroots organisers sought to supplement official datasets developed by the city of New York so as to better capture the experiences of those affected by the storm (for instance by carrying out additional, more detailed surveys).[50] Just as we saw in Chapters 3 and 5, in such cases citizens seek to foreground their own knowledge and experiences, overcoming the limitations of some aspects of technoscientific responses to disaster situations.

Box 6.6: What constitutes 'emergency'?

Elizabeth Ellcessor studies emergency media, and the ways in which tools such as emergency numbers or automated text alerts can be understood as mediators that help constitute what an 'emergency' is. This is important because, as she writes, emergency 'is defined – again and again – by not only those in crisis, but by emergency media workers who are themselves embedded in cultural contexts. Emergency and its mediation have served as sites of cultural power that protect some people while criminalizing or minimizing the plights of others'.[51] This means that emergency media are often predicated on a particular imagined norm: it can be hard for those with disabilities to access emergency services, for example, while in the United States (and beyond) whether one would call for the police in an emergency is conditioned by whether you are likely to receive help by doing so, or whether you might instead become the victim of racialised or class-based violence. Similarly, services such as extreme weather alerts, or texted instructions to stay at home during the pandemic, are triggered by particular individuals or groups deciding that the situation is, indeed, an emergency. The definition of 'emergencies' is therefore constituted by human choices – and occasionally errors, as when an alert about an incoming missile stating 'this is not a drill' was sent out to Hawaii residents in 2018.

Slow disasters and the climate crisis

The work described in the previous section complicates taken-for-granted understandings of the nature of disasters, drawing attention to how crisis events intersect with existing inequalities and to the ways in which disasters are in practice embedded in much longer timeframes than single moments of catastrophe. It is similarly exemplary of the ways in which technoscience and social and political processes intertwine. Many of the dynamics that this book explores – from negotiations of expertise to the presence of epistemic diversity – are foregrounded, made visible, in times of crisis. All of this has led to arguments for the value of reframing the nature of disasters. What if we thought of disasters as not necessarily 'fast' – sudden, isolated events – but rather as slow, unfolding over long periods and intersecting with existing injustices? What might this do to the politics of investment in disaster resilience, recovery, and preparedness, and of knowledge production around disasters and crises?

The notion of the slow disaster attempts to capture the kinds of harm that emerge not from spectacular moments of crisis, but from the insidious accumulation of damage. The 'slow disaster', Scott Gabriel Knowles writes:

> is a way to think about disasters not as discrete events but as long-term processes linked across time. The slow disaster stretches both back in time and forward across generations to indeterminate points, punctuated by moments we have traditionally conceptualized as 'disaster', but in fact claim much more life, health, and wealth across time than is generally calculated.[52]

To think of disasters as slow (even those that gain heightened visibility through particular events, such as Superstorm Sandy) is an intellectual move: it allows the acknowledgement of long histories of experiences of crisis, and the ways in which unequally distributed risks aggravate these. As such it has been used to analyse cases from water insecurity to pollution or healthcare. But it is also a means of engaging with the politics of harm and of disaster research. 'Fast' disasters are often publicly visible, prompting public responses through political attention, volunteering efforts, and fundraising. Understanding the nature of disasters in broader terms might direct attention to cumulative forms of harm, and to the ways in which disasters may be unfolding, unrecognised, even as we watch. Similarly, it might direct disaster research towards investigating the ways in which disasters are (as Knowles writes) *processes*, with long tails, structured by the intersections of diverse forms of harm and inequality. Perhaps it will become easier to act upon such harms if they are understood through the language of disaster.

The climate crisis is a central example of a slow disaster: one that unfolds gradually but catastrophically, and that is configured through long histories of specific intersections between technoscience and society. As Alice Bell charts, the 'greenhouse effect' and its impacts on global climate have a long history that is tied to the structure of industrial modernity, but also to the intersections of expertise, politics, and industry: as we saw earlier in this chapter, there were concerted efforts by those with a stake in fossil fuel industry to ensure that publics remained ignorant of, or at least confused by, the realities of the effects of fossil fuel use on the climate.[53] For Gabriel Knowles, understanding these histories is key to action in the present: 'formation of public policy that can meet the challenges of climate reality in the twenty-first century', he writes, 'relies on an ability to explain environmental change over long stretches of time, and to connect change to human actions'.[54] The long histories of the climate crisis structure what is think-able and do-able now, from a tradition of techno-optimism (the imagination that new technology, to be developed at some point in the future, will simply solve the problem) to the way in which scientists' historical wariness of engaging too much with the media has hampered the dissemination of climate knowledge.[55]

The ways in which climate science becomes public, and the effects that this has on public and political action on the climate crisis, are certainly too many and too complex to chart here.[56] But there are at least two central ideas that emerge from the scholarship discussed in this volume that seem important for engagement with the unfolding climate crisis. The first is that representations matter. As we saw in Chapter 4, the ways in which science and scientists are visible in public media help constitute the nature of science, and who 'belongs' within research. Public representations have public effects. The kinds of public accounts that are given of climate science and of climate change – both fictional and otherwise – will therefore come to shape what these things are. Erik Swyngedouw has analysed public discussions of climate change in the media and policy, arguing that while in the 2000s these stabilised to accept the reality of global warming (at least within mainstream media), they also took on apocalyptic overtones. Dramatic accounts (such as the 2004 film *The Day After Tomorrow*, in which a new, climate change-induced ice age grips North America over the course of a couple of days) as well as factual media present the climate crisis as a world-ending event, connoting 'an over-whelming, mind-boggling danger'.[57] This might, of course, be reasonable enough given the current science (though these representations tend to be of 'fast' disasters), but Swyngedouw is interested in what kind of space this creates for political engagement and action. His argument is that such apocalyptic visions actually close down action on the part of individuals and governments, effectively depoliticising climate change:

> At the symbolic level, apocalyptic imaginaries are extraordinarily powerful in disavowing or displacing social conflict and antagonisms … the presentation of climate change as a global humanitarian cause produces a thoroughly depoliticized imaginary, one that does not

> revolve around choosing one trajectory rather than another, one that is not articulated with specific political programs or socio-ecological project or revolutions.[58]

Presenting the climate crisis as world-ending may thus function in counter-intuitive ways, constituting it less as a problem to be acted on (through 'specific political programs') and more as an inevitable catastrophe – though also one that is inevitably and indefinitely being postponed, further negating the need for concrete action.

Box 6.7: The climate crisis as a justice issue

The impacts of the climate crisis may be accelerating, but it is in many ways a paradigmatic slow disaster in that its effects are differential. Those who are most vulnerable – such as the elderly or those living in low income countries – are affected most severely, are least responsible for it, and often have the least say in policy discussions.[59] At the 28th 'Conference of the Parties' (COP28, held in 2023) – the series of global meetings oriented to developing policy responses to climate change – collectives such as the Alliance of Small Island States criticised the decisions made as not being enough to protect those at risk, such as low-lying states.[60] At the same time, while there are increasing efforts to include marginalised communities into policy meetings such as the COPs, and to acknowledge the experiences and knowledges of indigenous peoples, these have been criticised as 'lip service'. On the one hand, writes Len Necefer, 'Indigenous knowledge tends to be localized and holistic, while western science's approach is highly specialized, and seeks universal truths. In many respects these different orientations can make threading these knowledge systems together fundamentally impossible'. On the other, 'the large brunt of the negative consequences of [climate change remediation] technologies will fall on marginalized communities who cannot afford to mitigate the impacts'. Ultimately, Necefer suggests that '[b]uilding equal relationships on the heels of hundred years of colonization is not a quick process, and time is running out'.[61]

The second central idea is that, if representations matter, so do local contexts, cultures, and meanings. The climate crisis is constituted through its public representations, but also through the ways in which these intersect with specific cultures and practices. This relates to the way in which technoscience is always made meaningful in particular contexts. 'Facts' may move through the world as mostly stable ideas, but they are constantly being made sense of through different frameworks, experiences, and priorities.[62] In the context of climate change, Candis Callison's book *How Climate Change Comes to*

Matter: The Communal Life of Facts offers an example of this.[63] Callison is exactly concerned with the ways in which climate science 'comes to matter' – is made meaningful – in different kinds of communities. She engages with five different (primarily US-based) communities, including science journalists, evangelical Christians, and Arctic indigenous representatives, looking at how groups in these communities oriented themselves to climate change and made sense of it through their particular languages, priorities, and ethics. For the Christian group that she worked with, for instance, 'concern for the environment is … a moral and biblical concern, hence the term creation care', while for the Inuit organisation climate change intersects with the struggle for self-determination and the drive to protect traditional lands.[64] Climate change, Callison suggests, is not a single thing: it is instead 'filled with meaning through its interaction with belief systems'.[65] This means that its effects, and action on it, cannot be discussed in universalising or totalising terms. Climate action cannot only be about the science, but about how this becomes paired with diverse cultures and with what Callison terms 'forms of life'. In this respect, it is as much a challenge for democracy and of engagement with different ways of making sense of the world as it is of public awareness or politics.

Conclusion

This chapter has surveyed the diverse ways that knowing and non-knowing are constituted, particularly in contexts of disaster and crisis. Just as in Chapter 5 we found differences between epistemic practices in different technoscientific and lay spaces, here we have seen that ignorance can itself be differently articulated and understood in different contexts. Rather than simply being an empty space or lack, it is socially constructed (with research and policy priorities taking knowledge production in some directions over others), and can take different forms (with known unknowns versus unknown unknowns being a particularly important distinction). The insidious effects of the formerly unknown unknowns that have emerged from technoscientific development – from CFCs to the green revolution – have led to heightened attention to managing ignorance, including through the use of risk as both a key technical measure and an idea that permeates modern societies. As a central framework risk has, however, been subject to criticism that it misses the much wider questions that publics often ask of technoscientific developments – not only whether they are safe, but whether they are desired, just, and socially valuable. Assessments of risk therefore cannot – or rather should not – be the only framework used in handling non-knowledge. Indeed, research into disasters and crises shows that affected communities are increasingly rejecting it as the sole means of understanding and making sense of these, for instance by developing their own datasets or by emphasising aspects of disasters such as their intersection with long-standing inequalities. Disasters are thus one space in which taken-for-granted assumptions regarding hierarchies of knowledge production are contested, and where public meanings regarding technoscientific issues may differ from those within science. Their framing is never simply a technical question, but relates to politics – as indeed we see in efforts to acknowledge the extended temporalities of crises by talking of slow disasters, and by highlighting the long historical trajectories of the climate crisis.

Many discussions around the intersections of technoscience with disasters and accidents thus relate to the question of whose knowledge counts, and of what form that knowledge should take. Tensions emerge around the role of expert knowledge and who can lay claim to this. In the next chapter we look at such expertise in more detail. How is it framed and understood, and is it under attack in contemporary society?

References

[a] Hess in Groß, M. and McGoey, L. (eds) (2015). *Routledge International Handbook of Ignorance Studies*. Routledge, p 142.

[1] There are unknown unknowns. (2024, 5 January). *Wikipedia*. https://en.wikipedia.org/w/index.php?title=There_are_unknown_unknowns&oldid=1193775042

[2] Saini, A. (2017). *Inferior: How Science Got Women Wrong-and the New Research That's Rewriting the Story*. Beacon Press. Williams, L.D.A. (2017). Getting Undone Technology Done: Global Techno-assemblage and the Value Chain of Invention. *Science, Technology and Society*, 22(1), 38–58.

[3] Hess, D.J. (2015). Undone Science and Social Movements: A Review and Typology, in Groß, M. and McGoey, L. (eds) *Routledge International Handbook of Ignorance Studies*, Routledge, pp 141–154, at p 142.

[4] D'Ignazio, C. and Klein, L.F. (2020). *Data Feminism*. MIT Press.

[5] Nielsen, K.H. and Sørensen, M.P. (2017). How to Take Non-knowledge Seriously, or 'the Unexpected Virtue of Ignorance'. *Public Understanding of Science*, 26(3), 385–392.

[6] Funtowicz, S.O. and Ravetz, J.R. (1993). Science for the Post-normal Age. *Futures*, 25(7), 739–755.

[7] Bocking, S. (2004). *Nature's Experts: Science, Politics, and the Environment*. Rutgers University Press, pp 114–115.

[8] Knorr-Cetina, K. (1999). *Epistemic Cultures: How the Sciences Make Knowledge*. Harvard University Press.

[9] Böschen, S., Kastenhofer, K., Rust, I., Soentgen, J., and Wehling, P. (2010). Scientific Nonknowledge and Its Political Dynamics: The Cases of Agri-Biotechnology and Mobile Phoning. *Science, Technology, & Human Values*, 35(6), 783–811, at p 790.

[10] Frickel, S., Gibbon, S., Howard, J., Kempner, J., Ottinger, G., and Hess, D.J. (2010). Undone Science: Charting Social Movement and Civil Society Challenges to Research Agenda Setting. *Science, Technology & Human Values*, 35(4), 444–473. Hudson, N. (2022). The Missed Disease? Endometriosis as an Example of 'Undone Science'. *Reproductive Biomedicine & Society Online*, 14, 20–27.

[11] In Kleinman, D.L. and Suryanarayanan, S. (2015). Ignorance and Industry: Agrichemicals and Honey Bee Deaths, in Groß, M. and McGoey, L. (eds) *Routledge International Handbook of Ignorance Studies*. Routledge, p 184.

[12] Huber, A. (2019). Hydropower in the Himalayan Hazardscape: Strategic Ignorance and the Production of Unequal Risk. *Water*, 11(3), Article 3, at p 15. https://doi.org/10.3390/w11030414

[13] McGoey, L. (2019). *The Unknowers: How Strategic Ignorance Rules the World*. Zed Books.

[14] Supran, G. and Oreskes, N. (2020). Addendum to 'Assessing ExxonMobil's climate change communications (1977–2014)' Supran and Oreskes (2017 *Environ. Res. Lett.* 12 084019). *Environmental Research Letters*, 15(11), 119401, at p 15.

[15] Stocking, S.H. and Holstein, L.W. (2015). Purveyors of Ignorance: Journalists as Agents in the Social Construction of Scientific Ignorance, in Groß, M. and McGoey, L. (eds) *Routledge International Handbook of Ignorance Studies*. Routledge, pp 105–113. Park, D.J. (2018). United States News Media and Climate Change in the Era of US President Trump. *Integrated Environmental Assessment and Management*, 14(2), 202–204.

[16] Mills, C.W. (2015). Global White Ignorance, in Groß, M. and McGoey, L. (eds) *Routledge International Handbook of Ignorance Studies*. Routledge, pp 36–46.

[17] Wekker, G. (2016). *White Innocence: Paradoxes of Colonialism and Race*. Duke University Press.

[18] Arora, S. and Stirling, A. (2023). Colonial Modernity and Sustainability Transitions: A Conceptualisation in Six Dimensions. *Environmental Innovation and Societal Transitions*, 48, 100733. https://doi.org/10.1016/j.eist.2023.100733

[19] Tousignant, N. (2018). *Edges of Exposure: Toxicology and the Problem of Capacity in Postcolonial Senegal*. Duke University Press, p 19.

[20] See Groß, M. and McGoey, L. (2015). Introduction, in Groß, M. and McGoey, L. (eds) *Routledge International Handbook of Ignorance Studies*. Routledge, pp 1–14.

[21] Michael, M. (1996). Knowing Ignorance and Ignoring Knowledge: Discourses of Ignorance in the Public Understanding of Science, in Irwin, A. and Wynne, B. (eds) *Misunderstanding Science? The Public Reconstruction of Science and Technology*. Cambridge University Press, pp 107–125, at 116.

22 Michael (1996).
23 Henwood, F., Wyatt, S., Hart, A., and Smith, J. (2003). 'Ignorance is Bliss Sometimes': Constraints on the Emergence of the 'Informed Patient' in the Changing Landscapes of Health Information. *Sociology of Health & Illness*, *25*(6), 589–607, at p 597.
24 Wehling, P. (2015). Fighting a Losing Battle? The Right Not to Know and the Dynamics of Biomedical Knowledge Production. In Groß, M. and McGoey, L. (eds) *Routledge International Handbook of Ignorance Studies*. Routledge, pp 206–214.
25 https://www.dictionary.com/e/misinformation-vs-disinformation-get-informed-on-the-difference/
26 Pennycook, G., Epstein, Z., Mosleh, M., Arechar, A.A., Eckles, D., and Rand, D.G. (2021). Shifting Attention to Accuracy Can Reduce Misinformation Online. *Nature*, 592(7855), 590–595, at p 590.
27 Nyhan, B. (2021). Why the Backfire Effect Does Not Explain the Durability of Political Misperceptions. *Proceedings of the National Academy of Sciences*, 118(15), e1912440117. https://doi.org/10.1073/pnas.1912440117
28 Altay, S. and Acerbi, A. (2023). People Believe Misinformation Is a Threat Because They Assume Others Are Gullible. *New Media & Society*, 14614448231153379. https://doi.org/10.1177/14614448231153379. Altay, S., Berriche, M., and Acerbi, A. (2023). Misinformation on Misinformation: Conceptual and Methodological Challenges. *Social Media + Society*, 9(1), 20563051221150412. https://doi.org/10.1177/20563051221150412
29 Lupton, D. (1999). *Risk*. Routledge.
30 Beck, U. (1992). *Risk Society: Towards a New Modernity*. SAGE, p 19, original emphasis.
31 Yeung, P. (2019, 29 May). The Toxic Effects of Electronic Waste in Accra, Ghana. *Bloomberg*. https://www.bloomberg.com/news/articles/2019-05-29/the-rich-world-s-electronic-waste-dumped-in-ghana
32 Akese, G.A. and Little, P.C. (2018). Electronic Waste and the Environmental Justice Challenge in Agbogbloshie. *Environmental Justice*, 11(2), 77–83, at pp 81–82.
33 Wynne, B. (2002). Risk and Environment as Legitimatory Discourses of Technology: Reflexivity Inside Out? *Current Sociology*, 50(3), 459–477.
34 Wynne (2002), p 472.
35 Sætra, H.S. and Danaher, J. (2023). Resolving the Battle of Short- vs. Long-term AI Risks. *AI and Ethics*. https://doi.org/10.1007/s43681-023-00336-y
36 Hanna, A. and Bender, E.M. (2023). AI Causes Real Harm. Let's Focus on That over the End-of-Humanity Hype. *Scientific American*. https://www-scientificamerican-com.uaccess.univie.ac.at/article/we-need-to-focus-on-ais-real-harms-not-imaginary-existential-risks/

[37] Whittaker, M. (2023, 11 June). Signal's Meredith Whittaker: 'These Are the People Who Could Actually Pause AI if They Wanted To'. *The Guardian*. https://www.theguardian.com/technology/2023/jun/11/signals-meredith-whittaker-these-are-the-people-who-could-actually-pause-ai-if-they-wanted-to

[38] Böschen, S., Kastenhofer, K., Rust, I., Soentgen, J., and Wehling, P. (2010). Scientific Nonknowledge and Its Political Dynamics: The Cases of Agri-Biotechnology and Mobile Phoning. *Science, Technology, & Human Values*, 35(6), 783–811.

[39] Lupton, D. and Tulloch, J. (2002). 'Risk is Part of Your Life': Risk Epistemologies Among a Group of Australians. *Sociology*, 36(2), 317–334.

[40] Slovic, P. (1987). Perception of Risk. *Science*, 236(4799), 280–285. Slovic, P., Finucane, M.L., Peters, E., and MacGregor, D.G. (2004). Risk as Analysis and Risk as Feelings: Some Thoughts about Affect, Reason, Risk, and Rationality. *Risk Analysis*, 24(2), 311–322.

[41] Lupton and Tulloch (2002), p 331.

[42] Jonas, H. (2014 [1973]). Technology and Responsibility: Reflections on the New Tasks of Ethics, in Sandler, R.L. (ed) *Ethics and Emerging Technologies*. Palgrave Macmillan, pp 37–47.

[43] Vaughan, D. (1990). Autonomy, Interdependence, and Social Control: NASA and the Space Shuttle Challenger. *Administrative Science Quarterly*, 35, 225–257, at p 226.

[44] Perrow, C. (1984). *Normal Accidents: Living with High-Risk Technologies*. Basic Books, p 4.

[45] Fortun, K., Knowles, S.G., Choi, V., Jobin, P., Matsumoto, M., de la Torre, P., et al (2017). Researching Disaster from an STS Perspective, in Felt, U., Fouché, R., Miller, C., and Smith-Doerr, L. (eds) *Handbook of Science and Technology Studies* (4th edn). MIT Press, pp 1003–1028.

[46] Liboiron, M. (2015). Disaster Data, Data Activism: Grassroots Responses to Representing Superstorm Sandy, in Leyda, J. and Negra, D. (eds) *Extreme Weather and Global Media*. Routledge, pp 144–162, at p 147.

[47] Liboiron (2015).

[48] Goldsmith, L., Raditz, V., and Méndez, M. (2021). Queer and Present Danger: Understanding the Disparate Impacts of Disasters on LGBTQ+ Communities. *Disasters*, 46(4), 946–973.

[49] Kera, D. (2010). Participatory Sensing after Fukushima: Fetish DIY Open Source Hardware for Community Science Projects. In *OZCHI 2011 Proceedings*. Canberra, Australia.

[50] Liboiron (2015).

[51] Ellcessor, E. (2022). *In Case of Emergency: How Technologies Mediate Crisis and Normalize Inequality*. New York University Press.

[52] Knowles, S.G. (2020). Slow Disaster in the Anthropocene: A Historian Witnesses Climate Change on the Korean Peninsula. *Daedalus*, 149(4), 192–206, at p 197.

[53] Bell, A. (2021). *Our Biggest Experiment: An Epic History of the Climate Crisis*. Catapult.

[54] Knowles (2020), p 203.

[55] Bell (2021).

[56] See this account for one discussion of what STS has to contribute to climate policy: Jasanoff, S. (2015). Science and Technology Studies, in Bäckstrand, K. and Lövbrand, E. (eds) *Research Handbook on Climate Governance*. Edward Elgar, pp 36–48.

[57] Swyngedouw, E. (2010). Apocalypse Forever? Post-political Populism and the Spectre of Climate Change. *Theory, Culture & Society*, 27(2–3), 213–232, at p 218.

[58] Swyngedouw (2010), p 219.

[59] Newell, P., Srivastava, S., Naess, L.O., Torres Contreras, G.A., and Price, R. (2021). Toward Transformative Climate Justice: An Emerging Research Agenda. *WIREs Climate Change*, 12(6), e733. https://doi.org/10.1002/wcc.733

[60] Rasmussen, A. (2023, 30 November to 13 December). *An Incremental Advance when Exponential Change is Needed: AOSIS Statement.* [Closing plenary]. COP28: United Nations Climate Change Conference, Dubai, UAE. https://www.aosis.org/cop28-closing-plenary-aosis-statement-on-gst-decision

[61] Necefer, L. (2022, 29 July). Paying Lip Service to Indigenous Knowledge Won't Fix Climate Change. *Outside*. https://www.outsideonline.com/culture/opinion/indigenous-knowledge-climate-change

[62] See Felt, U. and Davies, S.R. (eds) (2020). *Exploring Science Communication*. SAGE.

[63] Callison, C. (2014). *How Climate Change Comes to Matter: The Communal Life of Facts*. Duke University Press.

[64] Callison (2014), p 9.

[65] Callison (2014), p 247.

7

Experts and Expertise

The COVID-19 pandemic of the early 2020s marked a significant interruption into normal life for billions of people around the world, and cost millions of lives. Its impacts are still unfolding as I write, and touch everything from national economies to long-term healthcare needs and community cohesion.[1] But the pandemic is also illustrative in multiple ways of the themes discussed in this book. For months and years technoscientific research became highly visible, as governments referred to scientific advice to justify their decisions and as researchers explained their work in the media. As public audiences, we watched 'science in the making'* – unfinished, uncertain science – as scientists sought to make sense of the situation, offer the best advice possible, and develop vaccines and medical treatments. Given the emphasis on stable and uncontroversial knowledge within much science communication,[2] the pandemic was an almost unprecedented moment in which the uncertain, incremental nature of scientific research became public.† At the same time some researchers – like virologist Christian Drosten in Germany or epidemiologist Salim Abdool Karim in South Africa – became household names, shooting to a level of visibility no one would have anticipated.[3]

The story of one such researcher is particularly pertinent to the topics I will discuss in this chapter. Epidemiologist Neil Ferguson was a key figure in the UK government's pandemic response, a modeller whose work helped make the case for stringent lockdowns and physical distancing. But in early May 2020 he abruptly resigned from his government advisory role after newspaper reports that, during a period in which those in separate households were forbidden from intermingling, he had been visited twice by his married lover (or, as one report had it, 'trysting'). 'I accept I made an error of judgement and took the wrong course of action', Professor Ferguson

* The term comes from the work of Bruno Latour, among others.[a]

† This was a central challenge for policy making. While politicians emphasised that they were 'following the science', in practice this was often impossible, as the science was too uncertain to provide clear guidance or advice.[b]

said in a statement. 'I have therefore stepped back from my involvement in Sage (Scientific Advisory Group for Emergencies).'[4]

Why did Ferguson feel the need to resign? Perhaps it's not a question that needs asking – after all, his choice was treated as obvious at the time, with the UK health minister saying that it was the 'right decision' and that it was 'not possible' for him to continue in his expert advisory role. Other politicians pointed to the hypocrisy of Ferguson (and other high profile figures who broke lockdown restrictions), implying that his advice was tainted by his personal decision not to keep to the rules. While few people are sympathetic to hypocrisy (and while, in the UK context, politicians' own egregious breaking of lockdown regulations, and lies about doing so, later contributed to the fall of the prime minister[5]), it is worth reflecting on what this case can tell us about what we expect of experts. Ferguson's failure was not one relating to his knowledge or technical skill; he had made no errors in his modelling (or at least, this was not levelled as a criticism of him at this point). No one questioned his expertise, but his status as an expert still came into question. The case thus signals that we expect more of experts than their knowledge: their character and behaviour should be in line with the advice that they proffer. Ferguson failed in regard to this, rather than his technoscientific capacities in and of themselves being flawed.

This chapter explores this and other ambiguities associated with the notion of expertise. Ferguson's case is just one of many from the pandemic (and beyond) in which expert knowledge was contested or troubled. Indeed, the pandemic foregrounded the way in which expertise and experts have become central to contemporary societies, in times of crisis but also much more generally. Experts advise policy makers on key decisions, offer recommendations as to our lifestyles and behaviours, and make authoritative statements as to the nature of the universe and our place in it. The results of their epistemic practices are understood, as Steven Hilgartner writes, as being the 'gold standard' of knowledge production.[6] In this chapter we explore some characteristics of such expert knowledge, and how it relates to wider society, starting with the question of what expertise *is*. How has it been discussed and conceptualised? I answer this question by discussing two central frames for thinking about the nature of expertise: one in which it is a characteristic or attribute that is *possessed* or embodied by certain individuals, and another in which it is instead something that is, more or less successfully, *performed*.

Box 7.1: No one* is anti-science

Occasionally you will hear the critique that particular groups or communities are 'anti-science'. Anthony Fauci, the former director of the US Centers for Disease Control and

Prevention, has said, for instance, that '[t]here is a general anti-science, anti-authority, antivaccine feeling among some people in this country [the United States]'.[7] But when one examines the way in which such groups contest scientific advice, it seems that they are the opposite of anti-science: they mobilise research, researchers, and science to back up their claims.[8] They challenge particular scientific findings or advice, but simultaneously rely on the power of scientific authority. It is therefore not straightforward to distinguish between those who are 'pro' or 'anti' science. Rather, we might ask: what kind of knowledge is mobilised as evidence, on what grounds, and to what ends?

Expertise as possessed

When we think about the nature of expertise, the taken-for-granted model we tend to mobilise is that it is something that is possessed or embodied by particular people or groups. One *has* expertise, or one *is* an expert. And this is, indeed, a key way that expertise has been conceptualised. In this approach there are different ways of deciding who counts as an expert, and who does not. For some, academic credentials define who is an expert, with there being a hierarchy of expertise running from those with a university degree to researchers who are recognised as leaders in a particular field.[9] In this view an academic education (particularly gaining a PhD degree) and working in research are framed as key to expertise: one cannot be understood as expert without these formal accreditations.

But this is a rather limited perspective. If expert knowledge is 'exceptional knowledge, skills, and deliberative capacities',[10] or being a specialist of some kind,[11] then limiting its possession to people working in (or having credentials from) the academy seems unnecessarily constrained. There are numerous cases where useful knowledge turned out to reside not (only) in those with formal scientific training, but in individuals or groups with different kinds of 'lay', 'indigeneous', or 'local' knowledge. In the 1980s UK, for instance, government scientists gave confident advice to sheep farmers in uplands Cumbria about the effects of radioactive fallout from the Chernobyl nuclear reactor disaster. This was based on what turned out to be flawed technical models; as well as the presence of these flaws, they had ignored the extensive knowledge of the farmers themselves concerning the specificities of the geography of the region and the ways in which sheep behaved in it.[12] Or consider interactions between western science and indigenous knowledge. In a 2021 article, ecologists Christopher H. Trisos, Jess Auerbach, and Madhusudan Katti argue that the discipline needs to acknowledge its colonial

* Hardly anyone, other than in very special circumstances – such as rejecting modernity as a whole.

history, and the ways it had ignored or marginalised the deep knowledge and experience of indigenous communities, if it were to become a field that 'rejects extractive knowledge and instead generates knowledge that nurtures positive reciprocity with nature'.[13] Engagement with other forms of expert knowledge – those held by traditional communities – would render ecological research more robust. Or, finally, think of the role of patients and patient organisations in medical research. It is now commonly accepted that living with a particular condition (or caring for someone with one) provides experiences that are relevant to research into that condition, both in terms of priority setting and the substantive content of knowledge production.[14] It is thus now standard practice in many fields in health, social care, and public governance to draw on the knowledge of 'experts of experience'.[15]

To think of expertise as an attribute – something possessed by certain individuals or groups – thus does not mean that we have to limit it to those with formal credentials. What, though, is expertise composed of, and how do we recognise it when we see it? While, as noted earlier, an idea of 'specialist craft or knowledge [that] a person is said to possess' is generally central to definitions of expertise,[16] more recent accounts have emphasised that expert knowledge has multiple dimensions or qualities. Ideas of impartiality are often central: as Reiner Grundmann writes, 'experts are impartial which makes their advice trustworthy'.[17] In addition, for Ashley Rose Mehlenbacher, it is vital to think of expertise as comprised not just of knowledge but of how it is applied and used. Drawing on virtue ethics, she argues that experts should cultivate a kind of 'practical wisdom' that builds on experience and that is sensitive to the moral complexities of particular situations. Experts, she writes, 'are imbued with a kind of status that affords them both power and privilege … this privilege brings with it a moral duty to those with whom they have entered this social relationship'.[18] Thinking of expertise as solely concerned with facts and information is to ignore these social and moral dimensions and to invite distrust. Expertise should therefore be nurtured and developed with reference to character and wisdom as much as to specialist knowledge or skills in themselves. Similarly, Claudia Egher suggests that 'expertise is not solely a matter of intellectual and cognitive processes … affective labor plays an important part in its development, as it underlies people's efforts to coordinate with others'.[19] In this view expertise is multidimensional, incorporating knowledge and skills but also wisdom, relational capacities, and the ability to carry out appropriate emotional or affective work. Such perspectives on expertise can help us understand why Professor Ferguson – the epidemiologist whose misadventures I started the chapter with – was judged so harshly. His knowledge was not in question, but his character was: his actions signalled that he lacked the 'moral wisdom' that (good) experts should demonstrate, and seemed to disrespect the public audiences (sitting at home in lockdown) at whom his advice was directed.

Other discussions of the nature of expertise have focused on relevance, asking how one can, in contexts in which expertise can be found in different sites and kinds of people, assess what kinds of knowledge are relevant to particular topics or debates. Who should have a say, on what types of questions? Harry Collins and Rob Evans, for instance, provide a categorisation of expertise with the aim of helping to assess whether individuals should rightfully be involved in substantive discussion of particular questions. At its simplest, this schema includes (1) no expertise (no knowledge or experience of a domain); (2) interactional expertise, where one has some knowledge of a domain and can 'interact interestingly with participants' in it; and (3) contributory expertise, where one is able to make a substantive contribution to a field.[20] Importantly, they suggest that contributory expertise should *not* be understood as being confined to universities, professions, or other forms of accreditation: non-scientists may well have knowledge or experiences that allow them to contribute to knowledge production in a particular area (they cite the case of the sheep farmers mentioned earlier as an example of this). Indeed, those with interactional expertise – able to engage with and, to some extent, speak the language of, a particular domain – might be called on to mediate between such expert groups, who may not be aware of each other. On the other hand, those without expertise should not be invited to participate in debates within a field. As they write, theirs is a 'normative theory of expertise' that allows an analyst to differentiate between different forms of expertise and their suitability in the context of particular questions. Their schema can therefore be used to assess whose knowledge and expertise is relevant to specific debates – who should have a say, for instance, on particular environmental, health, or other technoscientific controversies.

Expertise as performed

While the various perspectives on the nature of expertise outlined above provide some insights into what expertise might be comprised of, in practice they are often challenging to mobilise as a means of differentiating between forms or degrees of expertise. In reaction to Collins and Evans' normative theory of expertise, for instance, many commentators pointed out that within controversies it is often exactly claims to expertise that are at stake, as actors seek to demonstrate the relevance of their knowledge or experiences – and the depth of their expertise with regard to it – to the issue in question. It is thus not straightforward to adjudicate on who has 'contributory' or 'interactional' expertise. Indeed, as Sheila Jasanoff points out, the answer to this question is context-dependent to the extent that who is able to credibly claim to be an 'expert' will look different within different national settings:

> [E]xpertise is not merely something that is in the heads and hands of skilled persons, constituted through their deep familiarity with the problem in question, but … is something acquired, and deployed, within particular historical, political, and cultural contexts. … Accordingly, who counts as an expert (and what counts as expertise) in UK environmental or public health controversies may not necessarily be who (or what) would count for the same purpose in Germany or India or the USA.[21]

This view takes us to a second key frame for thinking about expertise, in the form of approaches that explore how expertise is *achieved* in particular situations. In this view expertise is not something that one has, but that one performs – a 'practical achievement', as Claudia Egher puts it.[22] The central question for analysis and reflection is thus not who is an expert, but how certain actors become understood as expert in particular situations, and how this may be negotiated or contested. Who is perceived as an expert, in which spaces, and on what grounds?

Stephen Hilgartner is one key proponent of this 'dramaturgical' model of expertise. In a book titled *Science on Stage: Expert Advice as Public Drama*, Hilgartner discusses 'science advice as a form of drama' and examines 'how [this drama] is produced, performed, and subjected to critique'.[23] He uses the example of the US National Academy of Science, and in particular the production of a series of reports on diet and health in the 1980s, to explore how expertise can be (successfully) performed. Such performances are carefully crafted: advisors (individually or as expert groups) use rhetoric and particular forms of language to 'assert their credibility' – for instance, by emphasising their objectivity or distinguishing themselves from interest groups – but they also manage which information and interactions are visible on the 'front stage' (in public), as opposed to being in the 'back stage' (internal discussions). Back stages may be populated by many different actors and involve many different discussions, not all of which will be made public (though, as Hilgartner notes, the selective 'leaking' of key aspects of these discussions can be a highly effective strategy in tussles around credibility). Maintaining a particular persona (again, on an individual or collective level) on the front stage – for instance, that of independent, authoritative, objective expert (group) – is generally vital for the credibility of particular accounts, even where back-stage negotiations are much more complex or contested. Importantly, Hilgartner does not suggest that such performances are cynical or manipulative, but rather that they are an extension of the way in which we all manage the ways we present ourselves in everyday life. Drawing on the social theory of Erving Goffman, he argues that such stage and persona management is something that we all do in order to present the right kind of identity for particular situations (we generally act differently in home and work contexts, for example). Being an expert is thus in part a matter of how convincingly we can play this role. Formal credentials may help with this, but a successful performance is also dependent on what is expected of experts (objectivity and independence, for example), norms relating to particular contexts (in some places the expertise of lived experience may be prioritised over technoscientific knowledge), and how actors are able to meet these expectations.

This understanding of expertise does not help to make decisions about who has the 'right' kind of expertise with regard to specific questions, but it does allow us to analyse the dynamics of expertise in particular contexts, and to understand it as a flexible, relational category (meaning, who is understood as an expert depends on the context they are in, and the audiences or users they are relating to). In particular it signals that expertise is not something that is achieved once and for all, but that is dependent on specific contexts and situations. For Reiner Grundmann, for instance, '[e]xperts are primarily judged by clients, not necessarily by peers (professional or scientific); and they rely on trust by their clients'.[24] Status as an expert is thus largely dependent on whether one has (satisfied) users of one's expertise, and is relational in the sense that it cannot be achieved alone, but requires others to justify that status. (I may say I am an expert in mending 18th-century cuckoo clocks,* for instance, but unless others commission me to mend their clocks, and testify as to my excellence in doing so, then my claim is, if not entirely meaningless, at least easy to mock or contest.)

Box 7.2: 'The wrong science for the time and place'

It's 1830s Edinburgh, and a vicious competition is emerging around who will fill a vacant professorship in the prestigious domain of logic and metaphysics at the university.† The decision would be made by the town council, and so the candidates and their supporters staked their claims in public. On the one side was Sir William Hamilton, a rather staid philosopher with substantial support from traditional institutions such as the church. On the other was George Combe, a successful proponent of the then hugely popular (now discredited) science of phrenology. Both Hamilton and Combe actively sought to perform the role of an expert in a convincing manner, for instance by emphasising their credentials, the contexts in which they had previously worked, and the relevance of the knowledge that they produced. Hamilton got the job, but Thomas Gieryn argues that the final decision had nothing to do with the respective quality of the knowledge that he and Combe were producing. Rather, Combe and his science were

* For the record, I am not.

† This account is based on Thomas Gieryn's extensive research into the case.[c]

seen as too radical, too likely to shake up comfortable hierarchies within the city. His work was understood as impinging on the domain of the church, and – perhaps more importantly – as potentially leaching authority from the professional classes and other university professors. He wasn't the right kind of expert for this position, in this city: as Gieryn writes, he 'had the wrong science for the time and place'. In the end, Hamilton's performance of expertise was more compelling to the key audience at stake – the town council – than that of Combe.

Ideas of expertise as relational and performed have been taken up by other authors. In studying public debates on scientific topics, for example, Anne Kerr and her co-authors argue that speakers mobilised 'hybrid' subject positions: sometimes they presented themselves as scientific or technical experts, making reference to their knowledge and experience in specialist domains, while at others they spoke from lay perspectives, emphasising particular lived experiences or their status as citizens.[25] They moved between these personas depending on the discussion at hand, at times laying claim to these different forms of authority, Kerr and colleagues argue, 'almost simultaneously'. Similarly, Maria do Mar Pereira has studied the way in which scholars in women's, gender, and feminist studies in Portugal argue for the credibility of their discipline.[26] In a context in which such scholarship is often denigrated or viewed as a non-legitimate form of knowledge production, the researchers that she interviewed talked about needing to fit in with particular ideas about the nature of experts if they were to succeed. It helped, for instance, to present as a man, to dress well, or to have prestigious international experience.* Interviewees' positionality shaped whether they were understood as legitimate experts or not.

Viewing expertise as performance thus enables us to take it not as a given, but as something that is always unstable and open to question. Some performances are more durable than others: we can probably all think of contexts where expertise is readily accepted (many of us largely trust what accredited medical professionals tell us, for instance). But the value of this view of expertise is that it alerts us to the way in which expertise is situational. Who 'counts' as an expert may vary from context to context – and 'experts' themselves may draw on different kinds of knowledge to present themselves as authoritative in different situations.

* Of course, we can and should discuss whether these are helpful 'markers' of expertise. Pereira is not suggesting that this situation – in which dressing well, for instance, suggests legitimacy – is ideal, but rather describing how her interlocutors navigate a system in which their expertise is often denigrated.

Box 7.3: Performing the 'professor'

What do experts look like, and who is able to convincingly perform the role of someone with expert knowledge – a university teacher or professor, for instance? It has long been known that students assess the expertise of their teachers not only based on the knowledge that they convey, but on how they look and behave. Research into student evaluations shows that students may assess male- and female-identified teachers based on different criteria – for instance, as documented in one study, students expected 'female-identified teachers to engage in emotional labour to appear happy and relaxed' and gave poor evaluations when that wasn't the case.[27] Others describe the challenges of teaching when you don't have a 'normal professor body', understood as 'white, male, middle-class, middle-aged, able, heterosexual, and thin'.[28] Some bodies therefore have to work harder than others to be credited (and respected) as a teacher or professor.

Scientific expertise in the legal system

One place that performances of expertise, and assessments of these, come to matter is in the courtroom. Expert knowledge is now routinely mobilised within the legal system:* as two scholars of the relationship between law and science write, 'it is now expected that criminal trials will generally include some form of expert testimony'.[29] Such expert testimony is often related to forensic or DNA evidence, but other kinds of experts may be called on as well. One Science and Technology Studies professor, Simon Cole, tells the story of how he became an expert witness in one US legal case based on his research into the history of fingerprinting science and technology.[30] His knowledge in this area – he has written two influential books, *Suspect Identities: A History of Fingerprinting and Criminal Identification* and *Truth Machine: The Contentious History of DNA Fingerprinting* – meant that he was asked to testify as to the reliability of fingerprint identification in a particular case, but found his own expertise being called into question. The extract from the court transcript that follows amusingly shows how the judge, at least, was unfamiliar with (and perhaps suspicious of) his field of research. It opens with the defence attorney, Mr Zuss, asking Cole an opening question about his background:

* 'The' legal system? Much of the literature I am citing here comes from the United States. While there are similarities between legal systems and the practices that comprise them, it is important to note the limitations of the Anglophone literature here.

Q. [Mr Zuss]:	Good morning, Dr Cole. Dr Cole, please tell the Court of your educational background?
A. [Cole]:	Bachelor's degree from Princeton University in history and I have a PhD in science and technology studies from Cornell University.
Q.:	Could you tell the court –
The Court [Judge Brennan]:	I'm sorry, PhD in what?
The Witness [Cole]:	The field is called science and technology studies.
The Court:	Go ahead.

The question 'PhD in what?' suggests that Cole was not immediately convincing as a valid expert – and, indeed, his testimony was eventually ruled as inadmissible, as not relevant to the case in hand. In this case his knowledge – even when signalled by his connections to universities such as Princeton and his body of research – was judged as not expert (enough).

As well as illustrating how performances of expertise can be challenged, this case demonstrates one central way that the legal system becomes implicated in defining expertise, in the courtroom but also beyond. Unlike much scientific practice, the legal system is fundamentally oriented to making decisions and choices. Judgements – 'guilty' or 'not guilty', for instance – are clearly one aspect of this, but Cole's experiences point to another. In relying so frequently on expert input, courts must come to a decision concerning who classes as an expert, or of what knowledge is relevant to a particular issue, and they must do so within a particular timeframe. In adjudicating in this way their decisions have ramifications far beyond the courtroom or a specific case. Cole writes about the ways in which this one judgement – of the inadmissibility of his expert testimony – shaped how he was approached in the context of other legal cases, and his status as an 'expert' in public generally. Particularly in areas of controversial science, legal judgements about who is and isn't an expert can have far-reaching effects.

How are such decisions made? In the United States, the so-called 'Daubert Standard' gives a number of criteria for helping to decide what should class as admissible expert knowledge, including whether it has been peer-reviewed, whether it makes testable claims, and whether it is 'generally accepted' within a scientific community.[31] Quite apart from the flexibility of interpretation inherent to such criteria, however, it seems that judges also use a range of additional approaches and ideas when it comes to assessing expert claims. In studying decisions made in cases focusing on an area of controversial science in Finland, researcher Jaakko Taipale argues that judges engage in 'socio-technical review' to attribute or de-attribute credibility to particular experts.

This meant not only looking at the technical content of expert evidence, but relying on other indicators – such as how experienced experts were, the credentials they had, or their track record – to help assess expertise.[32] Similarly, Rees and White's account of a particular Canadian case (again in the context of a controversial medical diagnosis) indicates that experts who were seen as too interested – too committed to a particular position within a scientific debate – were looked on with disfavour. One expert witness, the judge wrote, 'was confusing his role as a witness, with that of an advocate'.[33] In the courts, as in other contexts, being able to signal objectivity and distance is thus essential to convincingly playing the role of expert.

Box 7.4: Bendectin and 'law-science knowledges'

The legal system is not only reliant on expert witnesses, but can function to shape both the content of scientific research and what comes to be accepted as consensus knowledge within it. In discussing 1980s and 1990s litigation around the anti-nausea drug known (in the United States) as Bendectin, Gary Edmond and David Mercer argue that controversy over its potential adverse effects was closed down not by the courts drawing on an external, objective body of stable scientific knowledge about it, but as scientific opinion developed in tandem with a series of legal cases. Scientists themselves disagreed about the drug's potential effects, as well as about the most appropriate methods and approaches to use to understand these. Eventual consensus about Bendectin's safety was thus achieved through what Edmond and Mercer call 'law-science knowledges – knowledges emanating from the specific interactions of law, science and society, and not from mythical images of extrinsic scientific assessment'.[34] In such cases litigation and other forms of legal action can serve to render visible disagreement within scientific communities, provoke specific kinds of research, and, eventually, force the closure of technoscientific controversies.

Expert knowledge and the legal system thus interact in interesting ways. Although there are often complaints that the two systems are radically different – that they have 'significantly different aims and normative commitments'[35] – it is clear that they are also mutually constitutive. Indeed, it is exactly the differences between the systems that can assist in their mutual shaping. The legal system is oriented to making decisions, and may therefore play a role in forcing the closure of controversies, including those involving science. As well as courts making decisions regarding who is an expert (as described earlier), Sheila Jasanoff has described how public and legal debates over issues such as the science of reproductive technologies, end of life care, and environmental harm both trigger new lines of research – those that can clarify the technoscientific questions at stake – and allow controversies to be resolved through the imposition of a binary decision: guilty or not guilty; yes or no; responsible or not.[36] Similarly, the law is itself now extensively populated by the use of scientific knowledge in creating particular forms of evidence and thus in shaping how its goal, that of justice, is achieved. Science defines the standards through which evidence is assessed, and, often, the expectations that juries and other participants have regarding it. Legal systems therefore prompt, assess, and shape research, while scientific expertise plays a central role in legal judgements and processes.

Box 7.5: CRISPR at the European Court of Justice

What counts as genetic modification? In Europe, where such techniques are closely regulated, this is a vital question, and one that has been subject to legal rulings at the national and international level. CRISPR is a gene-editing technique that allows for more precise modification of genomes than had previously been possible (as a signal of its significance, two scientists involved in developing it, Emmanuelle Charpentier and Jennifer Doudna, were awarded the Nobel Prize for Chemistry in 2020).[37] In 2018 the European Court of Justice ruled that plants created using CRISPR should be regarded as genetically modified organisms (GMOs), and labelled as such, despite calls from industry and scientific organisations to place them in a different category.[38] While the ruling was controversial, eliciting both critique and celebration, it demonstrates how the definitions of technoscientific terms or processes can be shaped by legal and regulatory actors. Like research itself, though, legal decisions may not be lasting: in 2024 new regulation was introduced that would ease the regulatory process for gene-edited crops, putting them into a category of plants created through 'New Genomic Techniques' that is distinct from other GMOs.[39]

A crisis of expertise?

The legal system is one site in which technoscientific expertise comes to matter, and one site where it is in turn given direction and focus. Beyond this one particular site expert knowledge is, of course, present in every chapter of this book: we find it mobilised in contexts of disaster (Chapter 6), contested and claimed through activism (Chapter 5), and subject to democratic deliberation (Chapters 2 and 8). The conceptualisations and models of expertise discussed in this chapter help us to understood how it is negotiated in such contexts, and its fluid and shifting nature. Based on such discussions we should therefore think of expertise as something rather complex, rather than as a fixed category or characteristic that can be straightforwardly attributed to a particular individual or community. As relational and performed, expertise can be rather fragile and precarious, not easily achieved or maintained (as the example of Neil Ferguson might illustrate).

Such a view of expertise is significant because to at least some commentators, contemporary societies are oriented towards it in new and important ways. There is a central tension or paradox at the heart of many modern societies: 'on the one hand', writes Claudia Egher,

> we are surrounded by more expertise because ever more domains of our lives have come under the authority of 'experts' and because expertise has been increasingly claimed by 'non-experts,' by people lacking official accreditations. On the other hand, the right and authority of experts to make decisions that impact the lives of many and the grounds upon which such decisions are made have been called into question, as the rise in anti-elitist and populist feelings over the last decade indicates.[40]

Expertise is thus a central ground for contestation and controversy, simultaneously central to life in contemporary 'knowledge societies', a source of authority and status that many seek to claim, and contested and politicised. Throughout the 2010s many countries saw events unfolding in which the views of accredited experts seemed to be disregarded: the exit of the UK from the European Union; the election of Donald Trump in the United States; widespread vaccine hesitancy around the world during the COVID-19 pandemic. Infamously, UK politician Michael Gove told an interviewer that 'I think the people in this country have had enough of experts' when commenting on the fact that few economists supported Brexit.* These developments have led to concerns about a 'crisis' in expertise, and a sense that expert knowledge is under attack and needs to be protected. Anxieties on the part of scientists and others about this crisis have in turn led to activism such as the March for Science[41] and to efforts to promote evidence-based decision making, on both individual (for instance with regard to climate behaviours) and systemic levels (in policy and politics). If (traditional) expertise is in crisis, then it is clear that it has its defenders as well as its attackers.†

---

* The full quote runs 'I think the people in this country have had enough of experts … from organisations with acronyms saying that they know what is best and getting it consistently wrong'. So the statement was, to some degree, qualified: it was experts from organisations with acronyms that were the problem.[d]

† In addition, the novelty of these developments should not be overstated. As we will see in Chapter 8, concerns about the relationship between science and politics have been present for many years. Indeed, Peter Weingart charts a central tension in late 20th-century

Is there a crisis of expertise? The answer, unsurprisingly, is complicated. On the one hand, trust in scientists as measured by surveys remains generally high,[42] and – as we saw in Chapter 5 – even activism and contestation around scientific issues tends to draw on scientific authority to make counter-claims or to support protest. On the other it is clear that there are now fewer spaces in which the authority of technoscientific expertise is uncritically accepted than there might have been in the past. Not only is there greater scope to question the positions of mainstream experts, in some communities it has become the norm to do so. What seems less clear is that it is technoscience in and of itself that is at stake in these contestations and controversies. For David Caudill, for instance, the 'culture wars' around science should be understood as clashes between different ideological systems or worldviews, each of which involves scientific knowledge but also beliefs and political commitments. '[I]deology is inevitable', he writes, such that scientists or other defenders of science must acknowledge that they 'cannot criticize others from a position of neutrality'.[43] Conflicts around expert knowledge should thus be understood as being between contrasting ideological systems – 'values, identities, and foundational commitments to a way of life'[44] – rather than attitudes to science per se. Similarly, what is known as 'science-related populism' – 'an antagonism between an (allegedly) virtuous ordinary people and an (allegedly) unvirtuous academic elite'[45] – is connected to other forms of populism, in that it signals discontent regarding a lack of agency and authority for ordinary citizens. What is at stake is less science itself, and more the political systems, positions, and spokespeople it is viewed as being connected to, and the ways in which these are viewed as threatening the sovereignty of 'the people'.

Vaccine hesitancy is illustrative of these complexities. Rather than such hesitancy being a blanket rejection of technoscientific advice, Maya J. Goldenberg emphasises the heterogeneity of reasons for and experiences of hesitancy, noting that there are a range of rationales for being concerned about vaccination and that these differ across different communities and around the world.[46] While some public discourse has (at least prior to the COVID-19 pandemic) critiqued concerns as 'affluenza' – the self-indulgent hesitations of middle-class liberals with no real experience of the horrors of infectious disease – Goldenberg shows that those in marginalised positions or who have negative experiences of healthcare institutions are (unsurprisingly) often hesitant with regard to the advice that such institutions give, including regarding vaccination.[47] Writing off all hesitancy as anti-science or as a

---

policy: on the one hand, there is 'scientification' of politics as science is used to inform policy; on the other, science is 'politicised' as it is used to legitimise policy decisions. Ultimately, 'given the legitimating function of (authoritative scientific) knowledge in politics, the general accessibility of that knowledge has led to a competition for expertise which intensifies controversies in policy-making rather than alleviating them'.[e]

purely political choice is thus overly simplistic. Focusing on low-income communities, she suggests that vaccine hesitancy may stem from a myriad of factors and is better understood as a 'crisis of trust' than a 'war on science'. In the context of childhood vaccination, it can signal:

> [A] political act that refuses community solidarity and rebuffs shared responsibility for public health, a suspicion of scientific and medical institutions that have participated in historical social injustices, a rejection of government intrusion on personal affairs, a reinstitution of family autonomy, a demand for less medical intervention and less corporate medicine (especially for children), and to some, a sign of good parenting.[48]

These are, Goldenberg writes, concerns about 'justice and values' that are expressed in and through vaccination choices, rather than solely relating to attitudes towards science. Similarly, studies of attitudes towards vaccination during the COVID-19 pandemic showed these to be tied to a range of factors, including the wider social context (public policy debate and infrastructure for vaccination, for instance) as well as the degree of trust that individuals have in the government. Importantly, it is clear that such choices were also shifting, developing as the situation changed such that vaccination 'decision-making is an ongoing process' that is shaped by people's 'social environment, the broader socio-political context they lived in, and people's relationship to political and scientific authorities'.[49]

Trust is thus a central theme in scholarship on attitudes to and uptake of vaccination. Trust in science is one aspect of this, but far more important is the way in which vaccination requires trust in the governments that promote it and the medical systems that deliver it.[50] As with other instantiations of the 'crisis of expertise', vaccine hesitancy thus encourages us to reject overarching narratives and to explore what is at stake within specific contexts of controversy or concern. As we have seen throughout this chapter, expertise (and responses to it) is not static but relational, dependent on the context in which it is performed. How convincing it is – and to what extent it will be trusted – will depend on many different factors, including experiences with the institutions in which expertise is embedded. Wholesale assessments of how expertise is encountered or responded to can only ever be provisional: the question of 'Who is an expert?' is always answered by 'It depends'.

Conclusion

In exploring the nature of expertise this chapter has not resulted in particularly clear answers regarding who has it. While a number of schemas and forms of categorisation exist, expertise can best be understood as something that is relational and shifting: what matters is whether you can give a convincing performance of being an expert. Indeed, given that expertise has different dimensions, and may rely on different types of experience, even technoscientific experts may at times draw upon other forms of knowledge or experience to render their performances more convincing. Whether one 'is' an expert therefore relies in part on the validation of others who credit you as such. We saw this dynamic in the context of the legal system, where expertise is simultaneously essential and always open to question, and where legal judgements are able to define whether someone has relevant technoscientific expertise. Expertise in turn plays a role in shaping society. As well as being central to court cases, assisting in the judgements made there, expert advice populates policy, medicine, and more (as we will see in more detail in the chapter that follows). Reports of the death of expertise are therefore exaggerated. While, with the rise of science-related populism (discussed in Chapter 1), we find more visible moments of resistance to or rejection of the advice of technoscientific experts, such dynamics should be understood as relating less to the idea of technoscientific expertise per se, and more to the contexts in which it is mobilised and people's experiences of these. As we also saw in Chapter 5 when discussing appropriation of scientific knowledge, technoscientific expertise generally remains perceived as a 'gold standard' that even critics of particular aspects of technoscience wish to lay claim to.

If this chapter has deconstructed the notion of expertise, and shown that it is contestable, situated, and created through social processes and judgements, that which follows explores how technoscientific expertise itself comes to play a role in society, and in particular in policy, politics, and democracy. How are the entanglements of technoscience and society constituted in the context of policy and questions of the public good?

References

[a] For instance, Latour, B. (1987). *Science in Action: How to Follow Scientists and Engineers through Society*. Harvard University Press.

[b] Evans, R. (2021). SAGE Advice and Political Decision-making: 'Following the Science' in Times of Epistemic Uncertainty. *Social Studies of Science*, 030631272110625. https://doi.org/10.1177/03063127211062586. MacAulay, M., Fafard, P., Cassola, A., and Palkovits, M. (2023). Analysing the 'Follow the Science' Rhetoric of Government Responses to COVID-19. *Policy & Politics*, 51(3), 466–485.

[c] Gieryn, T.F. (1999). *Cultural Boundaries of Science: Credibility on the Line*. University of Chicago Press.

[d] See Hawkins, A. (2016, 16 September). Has the Public Really Had Enough of Experts? *Full Fact*. https://fullfact.org/blog/2016/sep/has-public-really-had-enough-experts

[e] Weingart, P. (1999). Scientific Expertise and Political Accountability: Paradoxes of Science in Politics. *Science and Public Policy*, 26(3), 151–161.

[1] See, for instance, The British Academy (2021). *The COVID Decade: Understanding the Long-term Societal Impacts of COVID-19*. The British Academy. https://doi.org/10.5871/bac19stf/9780856726583.001

[2] Hine, A. and Medvecky, F. (2015). Unfinished Science in Museums: A Push for Critical Science Literacy. *Journal of Science Communication*, 14(2), A04. https://doi.org/10.22323/2.14020204

[3] Joubert, M., Guenther, L., Metcalfe, J., Riedlinger, M., Chakraborty, A., Gascoigne, T., et al (2023). 'Pandem-icons': Exploring the Characteristics of Highly Visible Scientists during the Covid-19 Pandemic. *Journal of Science Communication*, 22(1), A04. https://doi.org/10.22323/2.22010204

[4] BBC News (2020, 6 May). Coronavirus: Prof Neil Ferguson Quits Government Role after 'Undermining' Lockdown. https://www.bbc.com/news/uk-politics-52553229

[5] Bowman, D. and Roe-Crines, A.S. (2023). The End of the Rhetorical Line? The 'Partygate' Investigation into Former UK Prime Minister, Boris Johnson. *The Political Quarterly*, 94(3), 475–480.

[6] Hilgartner, S. (1990). The Dominant View of Popularization: Conceptual Problems, Political Uses. *Social Studies of Science*, 20(3), 519–539.

[7] CNN (2020, 28 June). The Situation Room. http://edition.cnn.com/TRANSCRIPTS/2006/28/sitroom.02.html

[8] Lynch, M. (2020). We Have Never Been Anti-Science: Reflections on Science Wars and Post-Truth. *Engaging Science, Technology, and Society*, 6, 49–57.

[9] Anderson, E. (2011). Democracy, Public Policy, and Lay Assessments of Scientific Testimony. *Episteme*, 8(2), 144–164.

[10] Mehlenbacher, A.R. (2022). *On Expertise: Cultivating Character, Goodwill, and Practical Wisdom*. Penn State Press, p 24.

[11] Grundmann, R. (2017). The Problem of Expertise in Knowledge Societies. *Minerva*, 55(1), 25–48.

[12] Wynne, B. (1992). Misunderstood Misunderstanding: Social Identities and Public Uptake of Science. *Public Understanding of Science*, 1(3), 281–304.

[13] Trisos, C.H., Auerbach, J., and Katti, M. (2021). Decoloniality and Anti-oppressive Practices for a More Ethical Ecology. *Nature Ecology & Evolution*, *5*(9), Article 9. See also discussion at: Lupin, D. (2024, 6 February). Decolonizing Science Means Taking Indigenous Knowledge Seriously. *This View of Life Magazine*. https://thisviewoflife.com/decolonizing-science-means-taking-indigenous-knowledge-seriously, which details some critiques of the article, including that it is important to allow indigenous groups to act as 'epistemic agents' on their own terms.

Decolonising science doesn't only mean making use of knowledge from indigenous communities, but allowed it be reshaped through it.

[14] See discussion in: Wilsdon, J., Wynne, B., and Stilgoe, J. (2005). *The Public Value of Science: Or How to Ensure that Science Really Matters*. Demos.

[15] Krick, E. (2022). Citizen Experts in Participatory Governance: Democratic and Epistemic Assets of Service User Involvement, Local Knowledge and Citizen Science. *Current Sociology*, 70(7), 994–1012.

[16] Grundmann (2017), p 26.

[17] Grundmann (2017), p 26.

[18] Mehlenbacher, A.R. (2022). *On Expertise: Cultivating Character, Goodwill, and Practical Wisdom*. Penn State Press, p 138.

[19] Egher, C. (2023). *Digital Healthcare and Expertise: Mental Health and New Knowledge Practices*. Springer Nature Singapore, p 15.

[20] Collins, H.M. and Evans, R. (2002). The Third Wave of Science Studies: Studies of Expertise and Experience. *Social Studies of Science*, 32(2), 235–296.

[21] Jasanoff, S. (2003). Breaking the Waves in Science Studies: Comment on H.M. Collins and Robert Evans, 'The Third Wave of Science Studies'. *Social Studies of Science*, 33(3), 389–400, at p 393.

[22] Egher (2023).

[23] Hilgartner (1990), p 6.

[24] Grundmann (2017), p 27.

[25] Kerr, A., Cunningham-Burley, S., and Tutton, R. (2007). Shifting Subject Positions: Experts and Lay People in Public Dialogue. *Social Studies of Science*, 37(3), 385–411.

[26] Pereira, M.M. (2018). Boundary-work that Does Not Work: Social Inequalities and the Non-performativity of Scientific Boundary-work. *Science, Technology, & Human Values*, 016224391879504.

[27] Adams, S., Bekker, S., Fan, Y., Gordon, T., and Shepherd, L.J. (2022). Gender Bias in Student Evaluations of Teaching: 'Punish[ing] Those Who Fail To Do Their Gender Right'. *Higher Education*, 83(4), 787–807, at p 803.

[28] Fisanick, C. (2007). 'They Are Weighted with Authority': Fat Female Professors in Academic and Popular Cultures. *Feminist Teacher*, 17(3), 237–255, at p 239.

[29] Rees, G. and White, D. (2023). Judging Post-Controversy Expertise: Judicial Discretion and Scientific Marginalisation in the Courtroom. *Science as Culture*, 32(1), 109–131, at p 109.

[30] Lynch, M. and Cole, S. (2005). Science and Technology Studies on Trial: Dilemmas of Expertise. *Social Studies of Science*, 35(2), 269–311.

[31] Caudill, D.S. (2020). Introducing Science and Technology Studies into the Expert Evidence Course, in Daly, Y., Gans, J., and Schwikkard, P. (eds) *Teaching Evidence Law: Contemporary Trends and Innovations* (1st edn). Routledge, pp 149–159.

[32] Taipale, J. (2019). Judges' Socio-technical Review of Contested Expertise. *Social Studies of Science*, 49(3), 310–332.

[33] Rees and White (2022).

[34] Edmond, G. and Mercer, D. (2000). Litigation Life: Law-Science Knowledge Construction in (Bendectin) Mass Toxic Tort Litigation. *Social Studies of Science*, 30(2), 265–316, at p 304.

[35] Jasanoff, S. (2006). Just Evidence: The Limits of Science in the Legal Process. *Journal of Law, Medicine & Ethics*, 34(2), 328–341, at p 329.

[36] Jasanoff, S. (1995). *Science at the Bar: Law, Science, and Technology in America.* Harvard University Press.

[37] Ledford, H. (2015). CRISPR, the Disruptor. *Nature*, 522(7554), 20–24.

[38] Gelinsky, E. and Hilbeck, A. (2018). European Court of Justice Ruling Regarding New Genetic Engineering Methods Scientifically Justified: A Commentary on the Biased Reporting about the Recent Ruling. *Environmental Sciences Europe*, 30(1), 52. https://doi.org/10.1186/s12302-018-0182-9. Stirling, A. (2018, 6 August). Is the New European Ruling on GM Techniques 'Anti-science'? *STEPS Centre.* https://steps-centre.org/blog/european-court-of-justice-ecj-gene-editing-anti-science/. Stokstad, E. (2018). European Court Ruling Raises Hurdles for CRISPR Crops. *Science.* https://doi.org/10.1126/science.aau8986

[39] Stokstad, E. (2024). European Parliament Votes to Ease Regulation of Gene-edited Crops. *Science.* https://doi.org/10.1126/science.zdjbra4

[40] Egher (2023), p 225. See also Grundmann (2017).

[41] Smith-Spark, L. and Hanna, J. (2017, 22 April). March for Science: Protesters Gather Worldwide to Support 'Evidence'. *CNN.* https://edition.cnn.com/2017/04/22/health/global-march-for-science/index.html. See also Riesch, H., Vrikki, P., Stephens, N., Lewis, J., and Martin, O. (2021). 'A Moment of Science, Please': Activism, Community, and Humor at the March for Science. *Bulletin of Science, Technology & Society*, 41(2–3), 46–57.

[42] Wellcome (2020). Wellcome Global Monitor 2020: Covid-19. *Wellcome Global Monitor.* https://wellcome.org/reports/wellcome-global-monitor-covid-19/2020

[43] Caudill, D.S. (2023). *Expertise in Crisis: The Ideological Contours of Public Scientific Controversies*. Policy Press, p 69.

[44] Caudill (2023), p 21.

[45] Mede, N.G. and Schäfer, M.S. (2020). Science-related Populism: Conceptualizing Populist Demands toward Science. *Public Understanding of Science*, 29(5), 473–491.

[46] Goldenberg, M.J. (2021). *Vaccine Hesitancy: Public Trust, Expertise, and the War on Science.* University of Pittsburgh Press.

[47] See also Prasad, A. (2022). Anti-science Misinformation and Conspiracies: COVID–19, Post-truth, and Science & Technology Studies (STS). *Science, Technology and Society*, 27(1), 88–112.

[48] Goldenberg (2021), pp 13–14.

[49] Zimmermann, B.M., Paul, K.T., Araújo, E.R., Buyx, A., Ferstl, S., Fiske, A., et al (2023). The Social and Socio-political Embeddedness of COVID-19 Vaccination Decision-making: A Five-country Qualitative Interview Study from Europe. *Vaccine*, 41(12), 2084–2092, at pp 2090–2091. See also: Paul, K.T., Zimmermann, B.M., Corsico, P., Fiske, A., Geiger, S., Johnson, S., et al (2022). Anticipating Hopes, Fears and Expectations towards COVID-19 Vaccines: A Qualitative Interview Study in Seven European Countries. *SSM – Qualitative Research in Health*, *2*, 100035. https://doi.org/10.1016/j.ssmqr.2021.100035

[50] Adhikari, B., Yeong Cheah, P., and von Seidlein, L. (2022). Trust is the Common Denominator for COVID-19 Vaccine Acceptance: A Literature Review. *Vaccine: X*, 12, 100213. https://doi.org/10.1016/j.jvacx.2022.100213

8

Science and Governance

What is necessary for building a well-functioning state? The answers, of course, are various, and are often enshrined in constitutions – the principles or laws that define and govern particular nations. In India one such constitutional principle is the notion of scientific temper, which is presented as one of the ten central duties of citizens. 'This clause', explain Anwesha Chakraborty and Poonam Pandey, 'makes embracing scientific and rational thinking and ways of life a duty and responsibility of Indian citizens'.[1] Based on the work and writings of India's first prime minister, Jawaharlal Nehru, scientific temper is:

> [T]he scientific approach, the adventurous and yet critical temper of science, the search for truth and new knowledge, the refusal to accept anything without testing and trial, the capacity to change previous conclusions in the face of new evidence, the reliance on observed fact and not on pre-conceived theory, the hard discipline of the mind – all this is necessary, not merely for the application of science but for life itself and the solution of its many problems.[2]

Indian citizens are thus expected to embrace not just science and its applications, but the mindset that is understood as tied to it: critical thinking, the capacity to change one's mind, the rejection of irrationality and 'religious temper' (which Nehru framed as the opposite of scientific temper). Science is thus constitutionally central to citizenship. Similarly, in 2009 the then-new President Obama promised to put science in its 'rightful place' in US society and politics. In both cases, the assumption is that science is central to society, the state, and to politics, and that its 'rightful place' is at the heart of the political system, speaking truth to power.[3] Indeed, for some commentators science and democracy are intimately connected, sharing central values such as rationality and working in constant support of one another.[4] In this view one cannot have one without the other.

The preceding chapter, in particular, raises some questions about what this means in practice. If expert knowledge exists within different groups and communities – not just those with formal accreditations – then how can it be effectively drawn upon within democratic decision making? If it is relational and contestable, who gets to decide who has relevant expertise with regard to particular policy questions? What, in practice, is its role in how (democratic) societies are governed? This chapter explores such questions by examining how technoscience is intertwined with policy, governance, and democracy. Importantly, we cover not just how scientific knowledge plays a role in policy making, but how technoscience is itself shaped through policy activities. As we see repeatedly throughout this book, science and technology never go untouched by the political and social processes they are utilised within. If they have a 'rightful place' in the state and in policy making, policy in turn also helps to constitute them.

Science for policy

Ideas of the 'rightful place' of science tend to assume a model of science and policy that is not dissimilar to the linear model discussed in Chapter 3: scientists produce knowledge, which is then used to inform policy decisions. Truth (scientific knowledge) is utilised by power (those in government). And indeed, experts are often called upon to offer advice to policy makers, politicians, and other decision makers. What do we know about the role of technoscientific knowledge in these kinds of policy processes?

We have already seen one example of expert advice in discussing Steven Hilgartner's *Science on Stage* in the previous chapter. Here Hilgartner describes one advisory mechanism in the US context, in which the National Academy of Sciences – a collective of renowned scholars – produces reports that seek to summarise or represent scientific consensus on particular issues and thus to inform policy discussions. Policy apparatuses look very different in different countries or regions, and handle expert advice differently – they have what has been described as different 'civic epistemologies', or logics of handling knowledge.[5] However, such expert reports or advisory groups are common in many political systems as a means of enabling 'evidence-based policy'.[6] As such, there have been efforts to define how scientists should think about their role in offering guidance to policy makers within such bodies and processes. Roger Pielke Jr's 2007 book *The Honest Broker* argues that there are (at least) four distinct roles that scientists might (individually or collectively) take in proffering advice to policy makers: they might act as 'pure scientists' who offer 'just the facts'; 'science arbiters' who answer fact-related questions; 'issue advocates' who argue for particular decisions or positions;

or 'honest brokers' who engage in 'decision-making by clarifying and, at times, seeking to expand the scope of choice available to decision-makers'.[7] These roles differ with regard to how engaged researchers are with policy, with the decisions at stake, and with wider public discussion and possibilities. Both 'pure scientists' and 'science arbiters' take a hands-off approach to policy, and assume that the two domains (of science and society or policy) are rather separate, while 'issue advocates' and 'honest brokers' are prepared to engage with political agendas, and with 'policy options rather than simply scientific results'. As the title of Pielke's book suggests, he is an admirer of those who act as honest brokers, but he is also clear that there can be a place for all of these ideal types of policy involvement, and that scientists should reflect on which is most appropriate within particular contexts. What is most important, he argues, is that advisors should acknowledge which role they are taking: often, scientists present themselves as 'pure scientists' engaged only with basic research while actually acting as 'stealth issue advocates' who are promoting a particular course of action. This conflation of specific policy choices with ostensibly objective or disengaged science is one reason that scientific advice may be viewed with distrust, or with the assumption that it is implicitly aligned with specific political positions (because it is).

Box 8.1: Against 'evidence-based' policy making

At first glance the notion of 'evidence-based' policy making – an idea that has been used 'as a battle-cry for public service reform' since the 1990s[8] – seems hard to argue with. Who wouldn't want public policy decisions to be based on existing knowledge? The term has, however, been subject to significant critique. One central concern is that its use is largely symbolic: while politicians and policy makers may say they want to base their decisions on evidence, studies show that its mobilisation often has little effect when prior partisan commitments are at stake. 'In Canada, for example', writes Joshua Newman, ' "tough on crime" legislation and mandatory minimum prison sentences continue to appear on the policy agenda, despite convincing evidence that these are expensive and ineffective ways to prevent crime'.[9] At the same time, others have shown that policy insistence on the use of scientific evidence actually closes down the possibility of diverse inputs into policy making. In an interview study with UK policy makers, Melanie Smallman finds that they mobilise what she calls a 'science to the rescue' imaginary, which assumes that technoscience is totally insulated from social and ethical values and judgements and that uncertainty can always be managed. This, she suggests, significantly limits the possibility of public knowledges or concerns being taken into account within policy. This in turn can lead to public dissent or controversy, as citizens seek to express their unease in the only avenues available to them. 'You might end up in the GM [genetic modification] situation', one of her interviewees, a civil servant says. '[T]he science becomes so dominant in the decision-making process around GM, that people are challenging the science, partly because they have other issues they want to express and there is no forum for them for doing that.'[10]

Pielke's book therefore offers guidance to researchers who wish to offer expertise to policy processes. As Pielke himself notes, though, it is not always straightforward to implement such advice, particularly if one is acting as an individual rather than as part of a committee or other collective.[11] Research that has studied science policy has shown how complicated science–policy interactions are, especially where there are key divergences not only around the science (where there may be ongoing uncertainty) but with regard to the values that lie behind a policy question. While scientists may hope to keep these separate, such distinctions are often not clear: as we saw in Chapter 2, research is by its very nature entangled with values, social processes, and judgements (such that the 'pure scientist' role is in practice almost impossible to inhabit, as Pielke notes). In addition, there are important differences between science and policy with regard to the logics and norms of how expert knowledge is conceptualised and used. Researchers often assume and hope that access to scientific research will resolve controversies by providing answers to central questions at stake – this, after all, is the assumption that lies behind the idea of 'evidence-based policy making'. But scientific controversies are often, at their heart, not actually about science, but particular positions or views. Abortion politics is only tangentially 'about' the science of reproduction; controversy over genetic modification is about more than the safety or otherwise of the technique. As Matthew Kearnes and colleagues write, in reflecting on a range of technoscientific controversies:

> For both GM and nuclear power, the social intensity of the arguments reflected not simply 'technical' issues held to be legitimate by governments and scientists, but also wider social relations in which the respective technologies were embedded (indeed, of which they were judged to be reflections) at their particular historical moments.[12]

It is therefore unclear that technoscientific knowledge actually does much to assist policy making on controversial topics, because the controversy emerges not from the science but the 'social relations' around it.

The quest to use science to end controversy is therefore often doomed. In fact, as science policy scholar Dan Sarewitz has argued, the paradox is that the addition of (more) scientific knowledge often makes controversies worse:

> In areas as diverse as climate change, nuclear waste disposal, endangered species and biodiversity, forest management, air and water pollution, and agricultural biotechnology, the growth of considerable bodies of scientific knowledge, created especially to resolve political dispute and enable effective decision making, has often been accompanied instead by growing political controversy and gridlock. Science typically lies at the center of the debate, where those who advocate some line of action are likely to claim a scientific justification for their position, while those opposing the action will either invoke scientific uncertainty or competing scientific results to support their opposition.[13]

Sarewitz's point is both that research is good at showing complexity and nuance (and therefore may not be well suited to answering the kinds of binary questions that policy making often involves), and that any particular scientific finding can usually be contested based on a critique of it, or on other research that has produced different findings. Even if a result or scientific argument is presented as definitive by those advocating for one position, it can be criticised or opposing findings offered by those committed to another. In addition, as we saw in the previous chapter, who or what has expertise on a particular question cannot be taken for granted. Performances of expertise are always subject to question, meaning that, in policy controversies, there may well be experts on both sides (any of whose expertise can be criticised by those on the other).

Making use of expertise within policy processes is therefore rarely straightforward. Even in relatively uncontroversial cases, scientific knowledge is entangled with values and assumptions, any of which may be questioned within political discussion and decision-making. Navigating the intersections between science and policy requires a nuanced approach – indeed, one of the challenges is that it is not always clear where technoscience ends and policy begins. Reuben Message has described the rise of 'policy-based evidence', where policy needs prompt particular kinds of research;[14] similarly, other studies describe the complex social interactions around ensuring that advice is taken up by those in policy, interactions that require not just expertise in and of itself but the ability to become a friendly 'insider' to policy.[15] In addition these interactions are not unidirectional. Even as expert advice is offered to and used by policy processes, policy is also shaping the kind of research that is carried out, and thereby the expertise and knowledge that is available. Policy decisions thus come to shape technoscience. The next section explores some of the ways in which they do so.

Box 8.2: Law, medicine, and rights in abortion

Much social research into abortion has been carried out in the highly polarised context of the United States. In contrast, one 2012 study examined how medical professionals talked about and negotiated the UK's legal framework, in which abortion is available up to 24 weeks with the agreement of two doctors. As Siân Beynon-Jones notes, this is indicative of a system in which the primary meaning of abortion is 'as an object of medical knowledge', and where 'doctors, rather than pregnant women, [are] the experts who must judge whether an abortion is necessary'.[16] Abortion later than the 24-week limit (initially defined based on ideas about the viability of the fetus outside the womb) is possible under special circumstances such as fetal abnormality. Beynon-Jones shows how the notion of 'viability' is contested, particularly with regard to the medicalised and technical way in which this is defined. Quoting Maureen McNeil, she writes that the notion of viability as '[t]he earliest gestation at which a baby has been born and has lived' (as framed by one of the medical professionals she interviewed) is 'an understanding of viability grounded in technical measurements of the biological functionality of a fetus rather than a pregnant woman's assessment of the "social sustainability of new life" '.[17] Similarly, the clause that allows abortions after 24 weeks in cases of 'physical or mental abnormalities' of the fetus has been criticised by disability rights activists, who argue that this 'represents discrimination against fetuses diagnosed with impairments and devalues disabled lives'. Thus even in a policy context in which abortion is framed as subject to medical judgement and permitted according to technical definitions of viability, such definitions are not clear-cut.

Policy for science

What forms does the co-constitution of technoscientific knowledge through policy and government take? First, there is the obvious aspect of regulatory control. Though science has a long history of self-governance, and though scientists have fought for the right to be independent of government or other interests and to maintain academic freedom, there are cases of legislation regarding the content of scientific research. This is most obvious where there are concerns regarding safety (such as nuclear physics research), or in the context of research that raises profound moral questions. The genetic engineering technique CRISPR-Cas9, for instance, rose to public prominence when it became clear that it offered a degree of precision that had not been possible with earlier methods and thus the potential to more accurately and effectively edit genetic sequences. In 2018 a Chinese scientist claimed to have used this technique to edit the genomes of human embryos, leading to the birth of twins and receiving widespread condemnation for acting unethically. Along with other forms of genetic engineering this technique is subject to specific regulations in a number of jurisdictions, effectively banning its use in contexts such as agriculture or human research.* Similar discussion has greeted the emergence of artificial intelligence (AI) technologies. At the time of writing AI regulation has been working its way through the European Commission and Parliament, and will likely result in European-wide legislation that regulates AI research and development based on the risks of potential applications.[18] The aim is to ensure that 'AI systems used in the EU are safe, transparent, traceable, non-discriminatory and environmentally friendly'.[19] The regulation will therefore guide AI research and development in both publicly funded research organisations and industry in some directions over others.†

While there are a number of such cases in which research is regulated in order to ensure public safety or the public good, such legislation remains unusual within science. This shouldn't suggest, however, that other research develops untouched by policy, politics, or public values. There are numerous subtle ways in which technoscience is guided by the involvement of societal factors or decisions. Sometimes this is done through what is known as 'governance': 'non-hierarchical, informal, network oriented, and cooperative forms of ruling such as soft law and self regulation … [where] the regulatory responsibility is shifted to the actors involved in the research, development, production, retail and disposal process'.[20] In contrast to formal regulation, governance thus involves informal activities or norms that nonetheless come to constrain how particular actors behave (voluntarily signing up to a code of conduct that dictates appropriate behaviour would be one example). In such cases researchers may develop their own forms of informal regulation, or align themselves around particular norms. Similarly, funders or governments may seek to encourage – but not enforce – certain kinds of research, both with regard to its focus and how it is carried out. Governance activities thus include codes of conduct or commitments to particular sets of principles: examples that have emerged in recent years include codes of conduct for research integrity,[21] moves to ensure the responsible use

* Such regulation tends, however, to be a moving target, and is constantly subject to updates and change. At the time of writing European regulation relating to gene-editing is in the process of (potentially) being updated.[a]

† It is not only regulation about the content of research and development that plays a role in shaping science. Some national systems have 'university laws' or similar legislation that define the role and place of academic institutions, and which can define the content of teaching or employment conditions for researchers. In Austria and Germany there are (as of 2023) heated debates concerning the limits that new legislation is putting on 'chains' of temporary employment at universities, effectively forcing researchers without permanent contracts to move elsewhere or to leave academia.

of research metrics (such as citation counts) such as the Declaration on Research Assessment,[22] guidance regarding how academic texts should be co-authored,[23] or ethical guidelines or principles to which researchers are asked to adhere.* Such codes or commitments are generally not enforceable, but rely on social pressure to bring about particular kinds of scholarship (those understood to be 'responsible' or 'ethical', for instance).

Funding and evaluation of research are also central to how technoscience develops. For example, the European Commission distributes billions of euros every year to researchers and industry using structures such as its framework programmes (the current one is called Horizon Europe, and lasts until 2027),[24] and the European Research Council, which funds 'excellent frontier research'.[25] The framework programmes, in particular, have clearly defined goals for the research that they fund – goals that might be oriented to climate change adaptation, for instance, or to the development of smart cities. In a context in which external funding is often necessary to gain the human or other resources needed to carry out research, if researchers want to acquire European funding then they need to be doing work that is in line with these goals or priorities, and proposing projects that relate to them. Europe is not unusual in this regard: put bluntly, funders' interests (whether those are public bodies or private foundations) shape the kind of research that is carried out. Most public research funders (at national as well as regional levels) choose to fund basic research as well as seeking to encourage particular forms of scholarship (those that they see as most useful) by asking researchers to submit particular kinds of project proposals. Indeed, there are whole areas of research that have emerged because of a specific need for them: forensic science, for instance, which has developed exactly to provide knowledge useful to the justice system. These dynamics sometimes lead to tensions and complaints regarding academic freedom versus demands for impact in particular areas, and discussion of the extent to which research should be guided by societal needs, values, and priorities, as opposed to scholars being able to follow what is of intellectual interest to them.[26]

Particular kinds of research, and ways of carrying out research, are also incentivised in other ways. In recent years evaluation and auditing have become central to academia in many research systems, and to how researchers are recruited, promoted, or rewarded. In this new 'audit culture'[27] the measurement of productivity is key: how much external funding has a researcher acquired? How many journal articles have they published, and how

* An entire regime now exists around the assessment of the ethics of research, from the use of 'Internal Review Boards' to guidance regarding principles such as informed consent, and can define what kind of research is do-able in very concrete ways. Such assessment is, of course, necessary – and has emerged out of horrific abuses of science such as the Tuskegee Study,[b] in which Black participants were left untreated for syphilis – but it is also important to acknowledge its role in shaping research.[c]

often are these cited? How do their students evaluate them? Such metrics may be used by universities and funders to assess researchers, for instance with regard to whether they should be granted a permanent position (gaining tenure, in the US system) or awarded research funding. While some form of assessment seems unproblematic – most forms of labour involve it – the manner it is which it has been carried out in academia has been heavily criticised. The Declaration on Research Assessment mentioned earlier rejects the use of quantitative metrics as the primary means of assessing the quality of scholars and scholarship, and argues for more nuanced approaches to how researchers are evaluated. Auditing publication metrics and grant income ignores other kinds of outputs from scholarly work, such as satisfied or inspired students, engagement with users or local stakeholders, or service to a disciplinary community. Relying on them forces researchers to prioritise particular aspects of science – those that will result in high profile publications – over others that might be equally valuable. Writing about the UK's national Research Assessment Exercise (RAE), Whitley et al note that:

> By the mid-1990s, RAE results were also being used increasingly as an input to university strategy, helping to decide which fields and departments to build up, and which departments to merge or close. … Department heads were explicitly tasked with improving RAE performance. Some adopted a more interventionist approach to managing faculty, advising them which research topics to pursue, whom to collaborate with, and in which journals to publish.[28]

Once again, choices and priorities on the part of those who fund research come to shape the content and process of scholarship in very concrete ways.

Box 8.3: Research evaluation around the world

Evaluation of research and researchers is now a central feature of global scholarship, in a context in which competition for resources – such as permanent positions in universities or funding for research projects – is intense even in wealthy nations. At stake is a fundamental question: what is 'high-quality' research, and how can it be measured and assessed? As described in the 2023 report 'The Future of Research Evaluation', metrics-based forms of evaluation are used in almost all research systems around the world.[29] At the University of Vienna (where I am based), for example, the university keeps track of how much money I apply for in project applications (and how many of these are successful); how many activities (such as conferences) I participate in or organise; and how many publications I produce, as well as the ranking of the journals that these appear in. At the same time the use of quantitative metrics to assess quality has come under increasing fire. The 2015 'Leiden Manifesto' notes that '[t]he problem is that evaluation is now led by the data rather than by judgement', and recommends, as its first principle, that '[q]uantitative evaluation should support qualitative, expert assessment'.[30] One reason for such critique is that these metrics are biased towards particular kinds of research and researchers. 'In excluding some forms of research and failing to harness a diversity of ideas', the Future of Research Evaluation report continues, 'there is a risk that current research assessment practices promote a mainstream/follower culture of dominant Western conceived models'.[31]

Responsibilisation and democratisation of technoscience

Technoscience is therefore shaped by policy processes and decisions in a number of ways. In this section I want to spend some time with a specific example of this, that of policy that seeks to align research with public values and needs. As discussed in Chapter 2, recent decades have seen a widespread sense that technoscience should be developed in a way that serves the public good. On the one hand this has involved the idea that technoscience should be 'responsibilised'; on the other, that it should be 'democratised'.

To talk of the need for responsible science and technology is to suggest that it is currently irresponsible, or at least not as responsible as it could be. Indeed, at least some such discussions can be rooted in the kinds of crises and technoscientific disasters discussed in Chapter 6, and in experiences of unintended effects of technoscientific development, from the toxic burden of microplastics to the long-term dangers of radioactive waste. The latter half of the 20th century saw, in many jurisdictions, a growing sense that such impacts should be anticipated as much as possible, and that technoscience should be encouraged to develop responsibly. This might mean integrating social, legal, and ethical analysis into it, and encouraging interdisciplinary reflection on the directions that research is taking.[32] It might involve reflecting on the trajectories of scholarship and technological development using futuring techniques (such as those described in Chapter 4) to consider what kinds of futures are being created through current research, and interrogating whether these are desirable.[33] But it might also involve critical engagement with how science is practised, and with the nature of good science. 'Responsibilisation' has thus also involved discussions of research integrity, and efforts to define good practice within research. As well as rejecting all forms of misconduct – most significantly, fabrication, falsification, and plagiarism – robust research should avoid 'questionable' research practices such as misusing statistics to inflate the significance of findings or selectively reporting results.[34] In this context responsibilisation is connected to concerns about the ways in which increasing pressures on researchers to publish (positive) results and to show success in their work are leading to low quality research that is, for instance, not reproducible by other scientists.[35] Responsibilisation can also mean openness, as in the movement for open science and open data, in which journal articles are freely accessible to anyone interested rather than being paywalled, and data that is produced by one group of researchers is available to others who may wish to re-analyse or otherwise reuse it.[36]

Box 8.4: Open science, FAIR, and CARE

'Openness' is increasingly a value that is promoted within technoscientific research. As Sabina Leonelli notes (and as we saw in Chapter 2), such promotion is in itself not surprising: the history of western science has been predicated on its activities being 'visible, intelligible, and receptive to critique'.[37] At the same time open science (for instance, in the form of freely available publications, 'open data' that is available for reuse or analysis by others, or 'open innovation' processes that allow citizens or users to participate in technoscientific development[38]) has been framed as a response to contemporary crises within research, including a 'reproducibility crisis' in which 'science faces widespread replication problems stemming from a mix of increased competition, incentives that reward improbable findings, and lax policing of bad or fraudulent science'.[39] One response has been the idea that greater transparency will help increase the robustness of research, with the 'FAIR principles' one concrete way that this is promoted. According to the principles, research data should be Findable, Accessible, Interoperable, and Reusable.[40] While the FAIR principles have rapidly become an important framework for research, and are promoted by science policy around the world, they have also been criticised.[41] One response to them is the CARE principles, developed by the Global Indigenous Data Alliance and standing for Collective Benefit, Authority to Control, Responsibility, and Ethics.[42] The development of the CARE principles sought to encourage 'open and other data movements to consider both people and purpose in their advocacy and pursuits', as well as the practical question of how to share scientific data.

Many of these moves come together in calls for responsible research and innovation (RRI), a movement that argues for 'a transparent, interactive process by which societal actors and innovators become mutually responsive to each other with a view to the (ethical) acceptability, sustainability and societal desirability of the innovation process and its marketable products'.[43] The aim is to ensure socially desirable research and innovation that are in line with public needs and desires. Given the non-trivial nature of ensuring this (particularly when it is not clear what the public good actually entails, and who should decide on this question), efforts to define RRI have included a number of different angles and practices. RRI should be anticipatory, involving reflection on where technoscientific development is taking us, as well as being open and inclusive (allowing perspectives from different stakeholders and groups within society).[44] Perhaps most importantly, it should be responsive, meaning that it should incorporate concrete mechanisms through which technoscience can respond to public concerns or values. Feedback, desires, and concerns from citizens and stakeholders should not be gathered without being used, or collected as an exercise in legitimation without any real effects, but integrated into the governance and directions of research and technology development. It is also significant that the notion of RRI incorporates not just research activities but innovation, something that aligns it with other 21st-century policy activities that seek to promote technology development. In Europe, there has been an 'increasingly singular emphasis on "innovation" as the solution to Europe's economic and social problems',[45] as Stevianna de Saille writes. Both in the European context and more broadly, RRI can be understood as a means of smoothing the way for such technology development.

RRI and other efforts to 'responsibilise' science thus overlap not only with each other, but with calls to open up or democratise technoscience. As we saw in Chapter 2, there have been both general calls to nurture deliberative and participatory forms of democracy, in which citizens can actively engage in discussions regarding political choices in formalised ways, and more specific experiments in rendering technoscience accessible and accountable to public groups. Arguments for this opening up have varied in their details, but often rest on the need for the involvement of diverse groups (beyond scientists and policy makers) in technoscientific research in contexts where there is the potential for significant impacts on wider society and a high degree of uncertainty and complexity. Daniel Fiorino, arguing against technocratic approaches to technoscientific risk assessment and management, suggests that there are at least three rationales for including lay citizens in technoscientific governance: a substantive argument that non-scientists have valuable contributions to make because they 'see problems, issues, and solutions that experts miss' and therefore add to the quality of research and decision making; a normative argument that, in democratic societies, inclusive deliberation and decision making is simply the right thing to do; and an instrumental argument that such inclusivity renders technoscientific decision making more legitimate and therefore more likely to be widely accepted.[46] Others have similarly suggested that the value of citizen participation in research and its governance can have multiple dimensions: on the one hand citizens may contribute their own knowledges and expertise on particular topics or questions (from scientific knowledge gleaned through citizen science to traditional knowledges or expertise based on experience[47]); on the other, even those without relevant expertise may offer 'perspectives and values' that emerge from diverse contexts, and that therefore assist in broad deliberation around technoscientific issues.[48] To repeat a point made in Chapter 2, the idea is that '[d]iversity results in better decision-making'.[49]

Organising public participation

The arguments and moves described earlier meant that, by the end of the 20th century, a critical mass of work in both science policy and Science and Technology Studies (STS) was calling for, and developing methods to use within, public participation in science. As with the science shops and formats for technology assessment described in Chapter 2, these forms of participation can be classified as elite-driven, or 'invited', as opposed to the uninvited forms of participation described in Chapter 5.[50] They were generally organised by academics, funders, or government bodies on topics of interest to them, rather than by laypeople or non-governmental organisations, with citizens then invited to participate in them. Experiments in participation mobilised a range of methods, often drawn from the deliberative democracy scholarship described earlier, and aimed in different ways to 'democratise' the governance and practice of technoscience.

One central format that became important in the spread of these experiments was the consensus conference. First developed in Denmark in the 1990s, this was a product of the Danish Board of Technology (DBT),[51] at that time a government-supported world leader in participatory approaches to science and technology. Consensus conferences, as the name suggests, seek to support deliberative discussions that could lead to consensus decisions on particular questions or issues in science policy.[52] Usually held over three days, they involve a representative panel of 14 citizens plus an 'expert panel', who give presentations and answer questions, and an organising group. The topic is defined in advance (for instance by a funder or by the organising group), and the citizens are asked to produce, by the end of the conference, a report that represents a consensual set of recommendations. Though originally developed in the Danish context (and embedded in specifically Danish assumptions, practices, and values concerning everyday democratic participation),[53] consensus conferences have now been held all over the world, on topics from nanotechnology to the future of automobility.

Citizens' juries are another frequently used deliberative format.[54] First developed in the United States as part of efforts to organise and institutionalise deliberative democracy,[55] the format was taken up and adapted to science-related topics. Like consensus conferences, they involve a panel of citizens, as well as a steering group who organise the process. In many citizens' juries, though, there have been efforts to put more power into the hands of those participating – to allow them to choose the topic that they focus on, for instance, and to select the experts who present evidence to them. The process again results in a report that makes recommendations, though the timing is somewhat different, with citizens' juries generally being spread out over a longer period of time. Differently again, scenario workshops are more explicitly future-oriented than consensus conferences or citizens' juries. This format (again first developed by the DBT) starts with a set of four scenarios (themselves developed by the organisers through a futuring process) that act as a starting point for deliberation between citizens and policy makers, experts, and representatives from industry.[56] The scenarios are critiqued and alternatives suggested, with a local plan of action being one possible output.

These three formats represent just a few of the vast array of participatory mechanisms that have now been developed and used in the context of technoscience (one 2005 review lists more than 100 types, though notes that this is non-comprehensive,[57] while the website and resource Participedia refers to 346 – though these are not limited to scientific topics[58]). All draw on deliberative democracy ideas of 'mini publics' – the bringing together of citizens from different backgrounds in a way that serves as a microcosm of the public sphere – and the value of discussion in reaching good decisions, though they implement these ideas in different ways. There have also been efforts to scale up such deliberation, and to go beyond decision making in specific local contexts. The 'World Wide Views' format[59] has been used to organise citizen panels around the world on topics such as climate and biodiversity, and involves groups of 100 citizens in different countries deliberating on the same topic, after which the results are compiled and presented to policy makers. The 2021 COP26 (Conference of the Parties) – the key venue for global policy debate about action on climate change – integrated results from a 'Global Assembly' in which 100 representative citizens from around the world deliberated on how humanity should address the climate crisis and made proposals on this.[60] And controversial areas of science such as gene editing continue to prompt national and international public consultations,[61] often run by dedicated organisations such as UK public participation charity Involve,[62] the DBT (which continues to be active in this space despite the loss of dedicated government funding), or the Spanish non-profit Deliberativa.[63]

There is therefore a rich landscape of ongoing activity oriented to consultation on and participation in technoscience and its governance – one that overlaps with a much wider set of activities that support deliberation, participation, and consultation in public decision making more generally. The rise of public engagement with science has often been framed as a success story for STS, one in which its 'normative turn' paid off and it was able to have a concrete impact on science policy and communication.[64] But the narrative I have outlined here – the arguments for such participation, and the different formats developed to support it – glosses over the anxiety and critique that often accompanied this work. Even as formats for participation were being developed and used, there was discussion of the ways in which they fell short of what was promised.[65] Methods such as consensus conferences are time- and resource-heavy, and their foci generally defined by powerful elites (such as governments or research funding bodies) rather than citizens. An emphasis on 'representativeness' in mini publics meant that individuals were reduced to specific identities, and called on to act as spokespersons of diverse communities that they might in practice only feel tangentially or partially attached to. Many of the recommendations and proposals generated were not listened to (and perhaps had never been intended to be listened to), generating frustration and 'participation fatigue'. And, as many people pointed out, the focus on deliberation could itself be limiting, excluding as it did forms of argumentation that go beyond talk, as well as the role of emotions and embodied experience.

There have been efforts to respond to these critiques, both in the context of technoscience and beyond. Participation is now more often co-created, rather than solely directed by elites. There have been experiments with longer-term deliberative processes, which engage communities over months or years rather than in one-off processes that may leave participants feeling frustrated or abandoned.[66] Similarly, some have experimented with formats that go beyond rarified conference table discussion to embed deliberation into particular environments, as well as acknowledging the importance of emotion and theatre in persuading fellow deliberators.[67] Participation has also been gamified in different ways, using card-based resources to stimulate discussion and to take it into more informal environments. Such experiments draw on similar critique and discussion in deliberative theory, which have resulted in a widening understanding of what deliberation is – not just reasoned argument that takes place at a particular moment, but spaces of interaction and reflection, distributed throughout society, which might include storytelling, art, personal testimony, humour, and embodied engagement with particular sites.

Box 8.5: Experimenting with participation

Public participation with technoscience takes many different forms. Increasingly, the distinction between 'invited' and 'uninvited' participation is becoming blurred, as participatory formats move away from highly structured 'top-down' mechanisms such as consensus conferences to more informal approaches. Public engagement may now take the form of hackathons,[68] participation in 'innovation hubs',[69] 'living labs' that act as real-world test spaces for innovations,[70] or co-creation activities in which citizens are invited to collaborate in knowledge production.[71] As this short list suggests, the current emphasis is often on participation in innovation, product development, and technological solutions to challenges such as urban infrastructure or climate change.

At the same time there are also efforts to give agency – and funding – to communities, rather than leaving control in the hands of funders or researchers. Programmes such as the Ideas Fund in the UK start with communities and their needs, enabling those groups to 'take the lead in working with researchers in innovative ways' rather than giving funding to researchers to incorporate citizens in their work.[72]

Expert knowledge and (deliberative) democracy

Public participation in science has therefore been one central focus for much science policy over the last decades, along with – and sometimes connected to – efforts to increase (responsible) innovation and technological development.[73] At the same time, there are some central ironies regarding how these formats have been developed and used, particularly with regard to their relation to the forms of activism and uninvited participation by publics in technoscience described in Chapter 5. As Ian Welsh and Brian Wynne write:

> [D]uring the same period that European and other states were increasingly inviting diverse publics into two-way public dialogue or engagement with science … publics as social movements engaged, uninvited, in the same range of issues, were being increasingly treated as threats to social and economic order, through their so-called anti-scientific behaviour.[74]

In other words, it was one thing for governments, funders, or researchers to invite select publics into participatory discussions on science, but quite another for those publics to get involved on their own terms. Consultation or carefully organised deliberation was acceptable, but activism or protest was (generally speaking) not, instead being framed as anti-scientific or as an invalid means of participation.

These tensions around how citizens are permitted or expected to engage with technoscientific governance relate to broader debates regarding the place of expert knowledge in democratic societies. Many theories of democracy, deliberative and otherwise, frame expertise as at least potentially a problem for democratic governance in that it involves specialist knowledge that is located in certain parts of society and not others. If a fundamental principle of democracy is that 'individuals are capable of self-determination and decision-making',[75] then expert activities are 'apparently out of the reach of democratic control … because of imbalances in knowledge'.[76] Expert knowledge can thus readily result in technocratic governance because it is assumed that expertise is in and of itself necessary for making good decisions about a particular field, and therefore that non-experts must be excluded from decision making. In the context of deliberative democracy, as political scientist Mark Brown notes, these issues have generally been side-stepped in that expert knowledge is framed as a resource that citizens can draw upon in making decisions, rather than being the subject of that deliberation.[77] In this view experts should ensure that their knowledge is accessible and comprehensible to citizens such that it can inform their discussions, feeding into deliberative processes that also involve the exchange of reasoned arguments about a topic or decision.

While such accessibility may well be a good thing, Brown argues that both this model, and the wider idea that expertise is a problem for democracy, are too simplistic. Framing expertise solely as a resource for deliberation ignores the ways in which it is likely to be located in a range of different actors, both from publics and research (as discussed in the previous chapter), and suggests sharp distinctions – between 'technical experts' and 'citizens' – that in reality do not exist. Similarly, it is not clear that one really needs to be a subject specialist in order to contribute to discussion of the trajectories and priorities of technoscientific research. Public scientific controversies are generally not about the content of facts – what Brian Wynne calls propositions about the world – but rather the meanings and purposes of technoscience and its relation to society. As Wynne writes, 'contestation is rarely only about propositional truths, but … about what is the proper public meaning and definition of the issue(s) being contested'.[78] Engaging with such meanings and definitions – questions such as: What exactly is the issue at stake? What is the purpose of technoscience? What role should it play in society or on particular questions? – does not require a detailed understanding of technical issues (what Collins and Evans would call, in the categorisation of expertise described in Chapter 7, contributory expertise) but much broader capacities, including the willingness to collectively reflect on meanings and values.

It is therefore not clear that technoscientific expertise is a special case for democracy, a field of activity that is somehow exempt or protected from public discussion and accountability. Indeed, many of the arguments for public deliberation and participation in technoscience exactly make the point that democracy is not well served by taking technoscientific knowledge as a protected realm. Not only does such knowledge present a limited view, one that (as we saw in Chapters 2 and 3) can embed racist, sexist, ableist and other assumptions and that can be as much subject to 'cognitive biases' as any other realm,[79] but in isolation it is not well-equipped to handle the complex and multidimensional challenges that face contemporary societies. '[M]ost prominent public issues today', Mark Brown writes, 'involve "ill-structured" or "wicked" problems that combine high decision stakes with a lack of societal agreement on both science and values'.[80] Insights from multiple domains, disciplines, and forms of knowledge – from citizens as well as from academia – will be necessary for dealing with such problems. It is for exactly these reasons that there have been calls for interdisciplinary research, 'extended peer review', and public participation. Rather than being reified, technoscientific expertise requires more opening up, more testing and expansion by those outside of the spaces in which it is produced. Given the immense power that a successful performance of expertise can achieve, '[e]xperts need to be kept in check, not given more power', writes Julian Reiss.[81] Technoscience is thus as much a part of democratic debate, and as valid a subject for deliberation and public discussion, as any other aspect of society.

Box 8.6: Whose democracy? Whose deliberation?

Much of the discussion in this chapter has taken for granted that there is something called 'democracy' that is clearly defined, stable, and universal. This is, of course, a radical simplification. Not only should we properly speak of 'democracies', to acknowledge that there are many versions of democracy, but there are many non-democratic contexts around the world in which technoscience is also present, contested, and subject to public participation. Rather little scholarship has thus far engaged with public engagement in such contexts.[82] Indeed, Linda Soneryd and Göran Sundqvist suggest that there may be particular affinities between STS research and deliberative democracy, because in both 'neither the authority of scientific knowledge nor the legitimacy of democratic governance are set once and for all', and thus frame both science and democracy as ongoing processes.[83] On the other hand, it is clear that regimes of all kinds are able to make use of the language of participation, citizen science, or public engagement. Hallam Stevens and Monamie Bhadra Haines describe how 'TraceTogether', an app developed by the Singapore government during the COVID-19 pandemic, builds 'on both the rhetoric and the technologies associated with citizen activism and citizen science', but directs this not towards citizen agency and mutual trust but to the reinforcement of state control. The app, they write, 'turns ideals born of dissent and protest on their head, using them to build trust not within a community but rather in government power and control'.[84] It is thus vital to interrogate what versions of 'democracy' are being realised in public participation tools and formats, and indeed whether this is even a meaningful category within particular situations.

Participatory technologies as constitutive

Thus far we have explored ways in which technoscience and policy interact, paying particular attention to moves to institutionalise public participation in scientific governance and to subject technoscience to deliberative discussion. These forms and formats of engagement have themselves now been subject to quite extensive study. What does such research tell us about their impacts and effects?

One important line of scholarship has taught us to understand participatory processes as themselves technologies that intervene in the world in particular ways.[85] As noted, scholarship has described how participation and deliberation can be used as tools for legitimation without democratisation, and has critiqued the limited impacts on policy or research that such processes have had.[86] More recent work has tended to explore what participation *does* do, albeit in less straightforward ways than observing whether it produces recommendations that are taken up by policy makers. In these accounts any form of participation should be understood as an intervention, something that works to constitute the worlds and entities it engages with. Participatory mechanisms are themselves *constitutive*, for instance of the issues they focus on or the publics they involve.*

This is perhaps most clear when one looks at the way in which lay participants have been framed within such processes. Alan Irwin, writing about a UK public consultation on the biosciences, noted that the 'institutional framing' of this consultation meant that 'participants appear as essentially reactive members of the public rather than as citizens in any more active sense of that term'.[87] In other words, even though the process may have been framed as an exercise in democracy, the version of citizenship it produced was highly limited, one that involved responding to the agenda of policy makers and scientists rather than more actively taking the lead in deliberation. Others have similarly pointed out that participatory methods constitute those doing the participating by giving them particular roles, or by imagining them in particular ways (as individuals ignorant of science and in need of educating, for example, or as potential consumers of technoscientific products). While citizens will not necessarily accept these roles – they may reject or subvert them – participation remains a moment in which categories such as citizenship, or stakeholders, or 'the public' are negotiated, and particular versions of them realised.[88] In the same way, participatory processes can help bring particular issues or areas of science into being. Holding a deliberative process on a particular topic (again, topics are generally selected by policy makers or others who fund or organise public

---

* This is part of a wider strand of work in STS that has argued that practices – including, but not limited to, participatory processes – are performative of the worlds and entities touched by them. Research methods have been a particular focus: when one studies an aspect of the world (for instance through carrying out surveys), that study is not simply describing an external, stable reality, but participating in constituting it. John Law's work on the 'Eurobarometer' survey instrument is one example.[d]

participation) reinforces its presence in the world, acting to stabilise what are perhaps nascent areas of research. Funding for deliberative processes centred on nanotechnology or synthetic biology during the 2010s could be viewed as an example of these dynamics. These areas of science were viewed (at least by many policy makers) as highly significant, and many promises were attached to them: nanotechnology would result in cures for cancer by 2015; synthetic biology would transform agriculture.[89] At the same there was little public awareness of them, and anxiety on the part of researchers and policy makers that public concerns might hinder their development. The wave of social research and participatory mechanisms dedicated to the fields could be understood as having stabilised them as discrete areas of scholarship, as well as serving – in many cases – to prepare the ground for their reception by framing participating publics as (potential) consumers.[90]

In contrast to what are perhaps taken-for-granted assumptions regarding public participation, then, such processes should not be understood as innocent or transparent formats that simply bring together pre-existing actors or issues. They will always play a role in constituting the aspects of technoscience that are their focus, as well as the various publics or groups that they seek to mobilise. Indeed, they can themselves be seen as technologies that embody certain assumptions and that travel (and are appropriated) in particular ways. As noted, the Danish Board of Technology has been a key actor in developing formats such as consensus conferences – formats that are now used all over the world. A number of scholars have discussed the ways in which the design of consensus conferences is related to the cultural context in which they emerged: the search for consensus as an ideal outcome, Maja Horst and Alan Irwin write, 'is deeply intermingled and embedded in the general political culture and context in Denmark'.[91] Designing public participation to have consensus as its goal makes sense in Denmark, in a culture in which citizens are used to thinking of political engagement as consensual in nature, but may not travel well to other contexts where politics works differently, or indeed where 'democracy' takes a very different form or is non-existent.* The way in which consensus conferences (and other 'mini public' formats) travel to different contexts, and are made meaningful to these, is therefore worthy of study in and of itself. Such travel also involves a central irony in that it requires the mobilisation of expertise *of* participation. In packaging deliberative formats for use in different places, and defining how they should be used, a 'new technocracy of procedure' has emerged,[92] raising questions and debate around who gets to decide how public participation

* The work of Monamie Bhadra Haines is one example of scholarship that has explored the different ways in which 'democracy', and science's relation to it, may be configured in contexts outside of the Global North.[e]

in technoscience should take place, and whether there is a need to 'open up' such decisions just as there was for other forms of scientific governance.

Finally, it is worth noting that in recent years (invited) public participation in technoscience has become increasingly entangled with (policy) assumptions regarding the importance of technological innovation, and framings of participation as assisting in this. As Pierre Delvenne and Hadrien Macq note, there has been 'a shift from participation in decision-making to participation in innovation-making'.[93] As participatory processes increasingly emphasise value-creation and the design of 'better' technologies over involvement in setting research agendas (Delvenne and Macq particularly mention newer formats such as hackathons and 'living labs' as implicated in this), they function to reinforce imaginations of contemporary societies as marked by inexorable technological development and, in particular, by close relations between business, technoscience, and economic growth. As well as constituting the entities involved in such forms of participation in particular ways (citizens as 'makers', for instance, or as users testing emergent technologies), public participation thus functions to create a politics of technoscience that is driven by commercial logics and actors. One of the things that is being made, along with particular technologies, is the very idea and realisation of 'innovation societies'.[94]

Conclusion

This chapter has explored some of the ways in which technoscience and policy and governance are entangled. Perhaps it is not surprising that a key space for the mutual shaping of society and technoscience is the realm of political decision making and government. Such decision making draws upon technoscientific expertise (through expert committees or other forms of advice), while simultaneously rendering technoscience subject to policy in the form of laws, informal guidance such as codes of conduct, funding regimes, and evaluation and assessment of research. Technoscientific scholarship is thus constantly guided through the actions and priorities of policy systems, which have, for instance, sought to bring about research that is responsible, directed towards a shared understanding of the public good, and innovation-oriented. In recent decades this has involved a particular interest in formalised mechanisms for public participation in science, for instance through consensus conferences or scenario workshops. Integrating these, the argument goes, will render research more robust (by drawing on relevant knowledge and experiences located in publics) as well as increasing the democratic legitimacy of decisions on technical issues. Importantly, these participatory mechanisms have both proven flawed in multiple ways – often leading to frustration or 'participation fatigue' – and are themselves technologies which embed particular assumptions and have particular effects. Just as with other technologies, it is worth scrutinising the nature of those assumptions and effects – for example, that citizens are consumers, that consensus is the correct outcome of democratic discussion, or that participation is about legitimating policy decisions rather than opening up new questions.

This chapter therefore makes the point, once again, that technoscience is always being shaped by social processes and decisions, while societal decisions are themselves fed by technoscientific expertise. Policy is a key site in which this takes place because, in many countries in the world, policy and government are central to imaginations of national identity and the nature of our collective lives – to who we are, as societies. The next chapter continues these reflections on the relation between technoscience and collective life by seeking to draw together the central ideas contained in the book as a whole, and relating these to questions of power, equity, and justice. How is technoscience implicated in power relations and in the maintenance (or erosion) of structural inequalities?

References

[a] Mir, T.G., Wani, A.K., Akhtar, N., and Shukla, S. (2022). CRISPR/Cas9: Regulations and Challenges for Law Enforcement to Combat its Dual-use. *Forensic Science International*, 334, 111274. https://doi.org/10.1016/j.forsciint.2022.111274. Stokstad, E. (2023, 7 July). European Commission Proposes Loosening Rules for Gene-edited Plants. *Science*. https://www.science.org/content/article/european-commission-proposes-loosening-rules-gene-edited-plants

[b] See Tuskegee University (nd). About the USPHS Syphilis Study. https://www.tuskegee.edu/about-us/centers-of-excellence/bioethics-center/about-the-usphs-syphilis-study

[c] See Hilgartner, S., Prainsack, B., and Hurlbut, J.B. (2017). Ethics as Governance in Genomics and Beyond, in Felt, U., Fouché, R., Miller, C.A., and Smith-Doerr, L. (eds) *The Handbook of Science and Technology Studies* (4th edn). MIT Press, pp 823–851.

[d] Law, J. (2009). Seeing Like a Survey. *Cultural Sociology*, 3(2), 239–256.

[e] Haines, M.B. (2020). (Nation) Building Civic Epistemologies around Nuclear Energy in India. *Journal of Responsible Innovation*, 7(sup1), 34–52.

[1] Chakraborty, A. and Pandey, P. (2024). Constructing 'Responsive' Publics: The Politics of Public Engagement under India's 'Decade of Innovation' Framework. *Asian Studies Review*, 48(1), 140–158, at p 153.

[2] Jawaharlal Nehru, quoted on p 52 of Mahanti, S. (2013). A Perspective on Scientific Temper in India. *Journal of Scientific Temper (JST)*, 1(1 & 2), Article 1 & 2. https://doi.org/10.56042/jst.v1i1

3. Hoppe, R. (1999). Policy Analysis, Science and Politics: From 'Speaking Truth to Power' to 'Making Sense Together'. *Science and Public Policy*, 26(3), 201–210.
4. Merton, R.K. (1973 [1942]). The Normative Structure of Science, in *The Sociology of Science: Theoretical and Empirical Investigations*. University of Chicago Press, pp 267–278.
5. Miller, C.A. (2008). Civic Epistemologies: Constituting Knowledge and Order in Political Communities. *Sociology Compass*, 2(6), 1896–1919. See also Jasanoff, S. (2005). *Designs on Nature: Science and Democracy in Europe and the United States*. Princeton University Press.
6. Jasanoff (2005). Cross, A. (2003). Drawing Up Guidelines for the Collection and Use of Expert Advice: The Experience of the European Commission. *Science and Public Policy*, 30(3), 189–192. Palmer, J., Owens, S., and Doubleday, R. (2019). Perfecting the 'Elevator Pitch'? Expert Advice as Locally-situated Boundary Work. *Science and Public Policy*, 46(2), 244–253. Khumalo, L., Crawley, K., Manyala, D., and Hassan, A. (2021). The Politics of Evidence in Parliamentary Oversight, in Khumalo, L., Morkel, C., Mapitsa, C.B., Engel, H., and Ali, A.J. (eds) *African Parliaments Volume 1: Evidence Systems for Governance and Development*. African Sun Media, pp 131–159.
7. Pielke, R.A. (2007). *The Honest Broker*, p 29.
8. Newman, J. (2017). Deconstructing the Debate over Evidence-based Policy. *Critical Policy Studies*, 11(2), 211–226, at p 211.
9. Newman (2017), pp 214–215.
10. Smallman, M. (2020). 'Nothing to do with the Science': How an Elite Sociotechnical Imaginary Cements Policy Resistance to Public Perspectives on Science and Technology through the Machinery of Government. *Social Studies of Science*, 50(4), 589–608, at p 597.
11. See also Oliver, K. and Cairney, P. (2019). The Dos and Don'ts of Influencing Policy: A Systematic Review of Advice to Academics. *Palgrave Communications*, 5(1), Article 1.
12. Kearnes, M., Grove-White, R., Macnaghten, P., Wilsdon, J., and Wynne, B. (2006). From Bio to Nano: Learning Lessons from the UK Agricultural Biotechnology Controversy. *Science as Culture*, 15, 291–307, at p 301.
13. Sarewitz, D. (2004). How Science Makes Environmental Controversies Worse. *Environmental Science & Policy*, 7(5), 385–403, at p 386.
14. Message, R. (nd). Policy-based Evidence Making. *Index of Evidence*. https://www.indexofevidence.org/policybased-evidence
15. Palmer et al (2019).
16. Beynon-Jones, S.M. (2012). Timing is Everything: The Demarcation of 'Later' Abortions in Scotland. *Social Studies of Science*, 42(1), 53–74, at pp 57, 59.
17. Beynon-Jones (2012), p 61.

[18] Veale, M. and Borgesius, F.Z. (2021). Demystifying the Draft EU Artificial Intelligence Act: Analysing the Good, the Bad, and the Unclear Elements of the Proposed Approach. *Computer Law Review International*, 22(4), 97–112.

[19] European Parliament (2023, 19 December). *EU AI Act: First Regulation on Artificial Intelligence*. https://www.europarl.europa.eu/news/en/headlines/society/20230601STO93804/eu-ai-act-first-regulation-on-artificial-intelligence

[20] Kurath, M. (2010). Nanotechnology Governance. *Science, Technology & Innovation Studies*, 5(2), 87–105, at p 88.

[21] For instance that developed by ALLEA, the European Federation of Academies of Sciences and Humanities: ALLEA – All European Academies (2023). *The European Code of Conduct for Research Integrity*. ALLEA – All European Academies.

[22] DORA (nd). The Declaration on Research Assessment. https://sfdora.org

[23] See ICMJE (nd). *Recommendations*. International Committee of Medical Journal Editors. https://www.icmje.org/recommendations

[24] European Commission (nd). *Horizon Europe*. European Union. https://research-and-innovation.ec.europa.eu/funding/funding-opportunities/funding-programmes-and-open-calls/horizon-europe_en

[25] European Research Council (nd). *ERC at a Glance*. European Union. https://erc.europa.eu/about-erc/erc-glance

[26] Holbrook, J.B. (2017). The Future of the Impact Agenda Depends on the Revaluation of Academic Freedom. *Palgrave Communications*, 3(1), Article 1.

[27] Shore, C. (2008). Audit Culture and Illiberal Governance: Universities and the Politics of Accountability. *Anthropological Theory*, 8(3), 278–298.

[28] Whitley, R., Gläser, J., and Engwall, L. (eds) (2010). *Reconfiguring Knowledge Production: Changing Authority Relationships in the Sciences and Their Consequences for Intellectual Innovation*. Oxford University Press.

[29] De Rijcke, S., Cosentino, C., Crewe, R., D'Ippoliti, C., Motala-Timol, S., Binti, A., et al (2023). *The Future of Research Evaluation: A Synthesis of Current Debates and Developments*. Centre for Science Futures.

[30] Hicks, D., Wouters, P., Waltman, L., de Rijcke, S., and Rafols, I. (2015). Bibliometrics: The Leiden Manifesto for Research Metrics. *Nature*, 520(7548), 429–431, at pp 429–430.

[31] De Rijcke et al (2023), p 8.

[32] Balmer, A., Calvert, J., Marris, C., Molyneux-Hodgson, S., Frow, E., Kearnes, M., et al (2015). Taking Roles in Interdisciplinary Collaborations: Reflections on Working in Post-ELSI Spaces in the UK Synthetic Biology Community. *Science & Technology Studies*, 28(3), 3–25.

[33] Selin, C. (2011). Negotiating Plausibility: Intervening in the Future of Nanotechnology. *Science and Engineering Ethics*, 17(4), 723–737.

[34] ALLEA (2023).

[35] See discussion of this movement within science, and how and why it has taken off as an area of concern, in Peterson, D. and Panofsky, A. (2023). Metascience as a Scientific Social Movement. *Minerva*, 61(2), 147–174.

[36] Beck, S., Bergenholtz, C., Bogers, M., Brasseur, T.-M., Conradsen, M.L., Di Marco, D., et al (2022). The Open Innovation in Science Research Field: A Collaborative Conceptualisation Approach. *Industry and Innovation*, 29(2), 136–185.

[37] Leonelli, S. (2023). *Philosophy of Open Science*. Cambridge University Press, p 1.

[38] Beck et al (2022).

[39] Peterson, D. and Panofsky, A. (2023). Metascience as a Scientific Social Movement. *Minerva*, 61(2), 147–174, at p 151.

[40] See GO FAIR (nd). *Go Fair Initiative*. https://www.go-fair.org/go-fair-initiative

[41] See, for instance, Bahlai, C., Bartlett, L., Burgio, K., Fournier, A., Keiser, C., Poisot, T., et al (2019). Open Science Isn't Always Open to All Scientists. *American Scientist*, 107(2), 78. https://doi.org/10.1511/2019.107.2.78. Carroll, S.R., Garba, I., Figueroa-Rodríguez, O.L., Holbrook, J., Lovett, R., Materechera, S., et al (2020). The CARE Principles for Indigenous Data Governance. *Data Science Journal*, 19(1). https://doi.org/10.5334/dsj-2020-043

[42] See Global Indigenous Data Alliance (nd). *CARE Principles for Indigenous Data Governance*. https://www.gida-global.org/care

[43] von Schomberg, R. (2011). Prospects for Technology Assessment in a Framework of Responsible Research and Innovation, in Dusseldorp, M. and Beecroft, R. (eds) *Technikfolgen abschätzen lehren: Bildungspotenziale transdisziplinärer Methoden*. Vs Verlag, pp 39–61, at p 39.

[44] Owen, R., Macnaghten, P., and Stilgoe, J. (2012). Responsible Research and Innovation: From Science in Society to Science for Society, with Society. *Science and Public Policy*, 39(6), 751–760.

[45] De Saille, S. (2015). Innovating Innovation Policy: The Emergence of 'Responsible Research and Innovation'. *Journal of Responsible Innovation*, 2(2), 152–168, at p 155.

[46] Fiorino, D.J. (1990). Citizen Participation and Environmental Risk: A Survey of Institutional Mechanisms. *Science Technology Human Values*, 15(2), 226–243.

[47] See examples in Chapters 5 and 6.

[48] Kinsella, W.J. (2004). Public Expertise: A Foundation for Citizen Participation in Energy and Environmental Decisions, in Depoe, S.P., Delicath, J.W., and Elsenbeer, M.-F.A. (eds) *Communication and Public Participation in Environmental Decision Making*. SUNY Press, pp 83–95. Krick, E. and Meriluoto, T. (2022). The Advent of the Citizen

Expert: Democratising or Pushing the Boundaries of Expertise? *Current Sociology*, 70(7), 967–973.

[49] Smith, G., Hughes, T., Adams, L., and Obijiaku, C. (eds) (2021). *Democracy in a Pandemic: Participation in Response to Crisis*. University of Westminster Press, p 4.

[50] Welsh, I. and Wynne, B. (2013). Science, Scientism and Imaginaries of Publics in the UK: Passive Objects, Incipient Threats. *Science as Culture*, 22(4), 540–566.

[51] https://tekno.dk/?lang=en

[52] Andersen, I.E. and Jaeger, B. (1999). Danish Participatory Models. Scenario Workshops and Consensus Conferences: Towards More Democratic Decision-Making. *Science and Public Policy*, 26(5), 331–340.

[53] Horst, M. and Irwin, A. (2010). Nations at Ease with Radical Knowledge: On Consensus, Consensusing and False Consensusness. *Social Studies of Science*, 40(1), 105–126.

[54] Wakeford, T. (2012). *Teach Yourself Citizens Juries: A Handbook*. DIY Jury Steering Group.

[55] Center for New Democratic Processes (nd). *How We Work*. Citizens Juries. https://www.cndp.us/about-us/how-we-work

[56] Andersen and Jaeger (1999).

[57] Rowe, G. and Frewer, L.J. (2005). A Typology of Public Engagement Mechanisms. *Science, Technology & Human Values*, 30(2), 251–290.

[58] Participedia (nd). *Home*. https://participedia.net

[59] World Wide Views (nd). *Home*. Danish Board of Technology. http://wwviews.org

[60] Global Assembly (nd). *Home*. https://globalassembly.org

[61] Dryzek, J.S., Nicol, D., Niemeyer, S., Pemberton, S., Curato, N., Bächtiger, A., et al. (2020). Global Citizen Deliberation on Genome Editing. *Science*, 369(6510), 1435–1437.

[62] Involve (nd). *Home*. The Involve Foundation. https://www.involve.org.uk

[63] Deliberativa (nd). *Home*. https://deliberativa.org/en

[64] Chilvers, J. and Kearnes, M. (2016). *Remaking Participation: Science, Environment and Emergent Publics*. Routledge.

[65] Delgado, A., Lein Kjolberg, K., and Wickson, F. (2011). Public Engagement Coming of Age: From Theory to Practice in STS Encounters with Nanotechnology. *Public Understanding of Science*, 20(6), 826–845.

[66] For instance, Powell, M.C. and Colin, M. (2009). Participatory Paradoxes: Facilitating Citizen Engagement in Science and Technology From the Top-Down? *Bulletin of Science, Technology & Society*, 29(4), 325–342.

[67] See Selin, C., Campbell Rawlings, K., de Ridder-Vignone, K., Sadowski, J., Altamirano Allende, C., Gano, G., et al (2017). Experiments in Engagement: Designing Public Engagement with Science and Technology for Capacity Building. *Public Understanding of Science*, 26(6), 634–649.

[68] Irani, L. (2015). Hackathons and the Making of Entrepreneurial Citizenship. *Science, Technology & Human Values*, 40(5), 799–824.

[69] Chakraborty, A. and Pandey, P. (2024). Constructing 'Responsive' Publics: The Politics of Public Engagement under India's 'Decade of Innovation' Framework. *Asian Studies Review*, 48(1), 140–158.

[70] Engels, F., Wentland, A., and Pfotenhauer, S.M. (2019). Testing Future Societies? Developing a Framework for Test Beds and Living Labs as Instruments of Innovation Governance. *Research Policy*, 48(9), 103826. https://doi.org/10.1016/j.respol.2019.103826

[71] Jansma, S.R., Dijkstra, A.M., and de Jong, M.D.T. (2021). Co-creation in Support of Responsible Research and Innovation: An Analysis of Three Stakeholder Workshops on Nanotechnology for Health. *Journal of Responsible Innovation*, 1–21. https://doi.org/10.1080/23299460.2021.1994195

[72] Ideas Fund (2023). Supporting Community-led Collaboration with Researchers: An Insight Report. https://readymag.com/theliminalspace/4230631/

[73] De Saille (2015).

[74] Welsh and Wynne (2013), p 541.

[75] Baghramian, M. and Martini, C. (2022). *Questioning Experts and Expertise.* Taylor & Francis, p 2.

[76] Turner, S. (2001). What is the Problem with Experts? *Social Studies of Science*, 31(1), 123–149, at p 123.

[77] Brown, M. (2014). Expertise and Deliberative Democracy, in Elstub, S. and McLaverty, P. (eds) *Deliberative Democracy: Issues and Cases.* Edinburgh University Press, pp 50–68.

[78] Wynne, B. (2003). Seasick on the Third Wave? Subverting the Hegemony of Propositionalism: Response to Collins & Evans (2002). *Social Studies of Science*, 33(3), 401–417, at p 404.

[79] Baghramian and Martini (2022).

[80] Brown (2014), p 62.

[81] Reiss, J. (2019). Expertise, Agreement, and the Nature of Social Scientific Facts or: Against Epistocracy. *Social Epistemology*, 33(2), 183–192, at p 191.

[82] Harris, J. (2020). Science and Democracy Reconsidered. *Engaging Science, Technology, and Society*, 6, 102–110.

[83] Soneryd, L. and Sundqvist, G. (2023). *Science and Democracy: A Science and Technology Studies Approach.* Bristol University Press, p 15.

[84] Stevens, H. and Haines, M.B. (2020). TraceTogether: Pandemic Response, Democracy, and Technology. *East Asian Science, Technology and Society: An International Journal*, 14(3), 523–532, at pp 524–525.

[85] See, in particular, Chilvers and Kearnes (2016).

[86] Hagendijk, R. and Irwin, A. (2006). Public Deliberation and Governance: Engaging with Science and Technology in Contemporary Europe. *Minerva*, 44(2), 167–184.

[87] Irwin, A. (2001). Constructing the Scientific Citizen: Science and Democracy in the Biosciences. *Public Understanding of Science*, 10(1), 1–18.

[88] Felt, U. and Fochler, M. (2010). Machineries for Making Publics: Inscribing and De-scribing Publics in Public Engagement. *Minerva*, 48(3), 219–238. Goven, J. (2006). Processes of Inclusion, Cultures of Calculation, Structures of Power: Scientific Citizenship and the Royal Commission on Genetic Modification. *Science, Technology & Human Values*, 31(5), 565–598.

[89] See discussions in Corner, A. and Pidgeon, N. (2013). Nanotechnologies and Upstream Public Engagement, in Harthorn, B.H. and Mohr, J.W. (eds) *The Social Life of Nanotechnology*. Routledge, pp 169–194.

Marris, C. (2014). The Construction of Imaginaries of the Public as a Threat to Synthetic Biology. *Science as Culture*, 24(1), 83–98.

[90] See, for instance, Marris, C. and Calvert, J. (2019). Science and Technology Studies in Policy: The UK Synthetic Biology Roadmap. *Science, Technology, & Human Values*, 45(1), 34–61.

[91] Horst and Irwin, A. (2010), p 116.

[92] Voß, J.-P. and Amelung, N. (2016). Innovating Public Participation Methods: Technoscientization and Reflexive Engagement. *Social Studies of Science*, 46(5), 749–772.

[93] Delvenne, P. and Macq, H. (2019). Breaking Bad with the Participatory Turn? Accelerating Time and Intensifying Value in Participatory Experiments. *Science as Culture*, 29(2), 245–268, at p 247.

[94] See also De Saille (2015).

9

Technoscience, Power, and Justice

This chapter is slightly different to those that have preceded it. As I worked on the chapters you have read so far, I increasingly felt that it was important to contextualise the discussions they contain (and the spaces that they describe) through a more explicit engagement with power. Indeed, I came to see this as even more urgent than thinking about technoscience in the context of democracy – the kinds of debates discussed in the preceding chapter. It is certainly important to consider the place of technoscience in democratic societies, and the ways in which it can be subject to deliberation and debate, but at the same time democracy is a slippery concept, and one that looks very different in different contexts. Many countries, spaces, and processes are not committed to the version of democracy that is celebrated in deliberative theory. It is therefore necessary to find ways of critically reflecting on technoscience and its place in collective life that do not simply end at the idea of democratisation. This chapter uses scholarship concerned with power and justice to do this, returning to many of the sites and processes I have discussed so far to consider their intersection with questions of equity[1] – by which I mean questions of fairness and equitable access to the opportunities and benefits of contemporary societies.*

Why should we critically interrogate technoscience's entanglements with collective life in this way, and reflect on how it relates to power and structural inequalities? The answer to this question is for me exemplified by a moment at a science communication conference I attended in 2023. Like many other fields, science communication has, over the course of the last years, started to reckon with the ways in which it continues to incorporate institutionalised racism, sexism, and other forms of oppression. I have already mentioned

* Equity is different to equality: as Max Liboiron and colleagues write, '[e]quality involves treating everyone exactly the same, and as a result has no impact on the uneven positions from which different people start. Equity, in contrast, is sensitive to the different positions of participants and so is potentially transformative of power relations'.[a]

(in Chapter 5) analyses of science communication that show the ways in which it is organised around Whiteness and middle-class values;[2] other work has sought to explore what it might mean to queer, decolonise, or otherwise diversify science communication practice.[3] Both the urgency and the sensitive nature of these moves became clear at the conference. Science has a history of oppressing Black and queer bodies, one Black participant said. How can it be held accountable for this? Similarly, it has a history of ignoring women's pain, and of framing their bodies as deviations from a male norm.[4] Why should those of us who inhabit bodies subject to such forms of oppression or ignorance trust or celebrate it?

Technoscience is a hugely powerful tool – a golem created by humans in their service, as one 1990s book put it.[5] It provides knowledge about the world in a manner that is reliable and robust, and offers insights that are, to many, thrilling, awe-inducing, practically useful, and endlessly fascinating. But we should not ignore the ways in which it is and has been implicated in maintaining oppression and inequality, nor those in which it has caused harm. If we are to think about its place in contemporary societies, and to consider how it is shaped by and in turn shapes those societies, then it is vital to engage with how it is implicated in power relations and questions of justice and (in)equity.

This chapter therefore considers research that has explored these dynamics. Because this discussion draws on many of the ideas we have already encountered in the preceding chapters, I start by discussing some cross-cutting themes and ideas that have repeatedly emerged as we have considered spaces and processes through which technoscience and society are mutually constituted.

Key themes and ideas

A first central idea is the one that I have used to frame the book as a whole: that *technoscience and society are not separate entities or domains, but are always entangled with each other.* Despite the prevalence of taken-for-granted assumptions or metaphors (such as the ivory tower, discussed in Chapter 2) that emphasise the separation between scientific practice and social or societal processes and values, we have repeatedly seen that technoscience is constantly inflected by social processes, and that contemporary societies operate with and through technoscientific ideas, products, and forms of assessment. We have observed, for example, the ways in which particular values and choices are embedded in science and technology (Chapters 2 and 3), how activism around technoscientific issues may involve a rejection of those implicit values (Chapters 3 and 5), how technical models, promises, and visions direct societies along particular trajectories (Chapter 4), and the ways in which expert advice is central to policy and legal decision making

(Chapters 7 and 8). Importantly, the mutual constitution of technoscience and society is exactly mutual, operating in both directions. Technoscience shapes society, but society also shapes technoscience. The ways in which these interactions occur are not uniform; what is most productive, then, is to explore the ways in which the social and the technoscientific take form within specific contexts and sites.

One aspect of these processes of mutual constitution is that *representations matter*. That is to say, it is not only specific physical products or social interactions relating to technoscience that are important in its shaping and effects, but the ways in which technoscience and society are talked about, represented, and imagined. Metaphors, we saw in Chapter 2, help reinforce a vision of science and society as separate domains, while the ways in which science and scientists are depicted in public media may reproduce stereotypes concerning who belongs in research (Chapter 4). Similarly, the ways in which climate science is represented can shape how the climate crisis becomes think-able and amenable to political action (Chapter 6). Public and popular media – and specifically how technoscience comes to be represented in particular ways within these – are thus one space in which technoscience and society are constituted.

A third idea relates to another way in which technoscience becomes particularly visible in wider society. Technoscientific knowledge is widely understood as *expert* knowledge – as uniquely authoritative, reliable, the 'gold standard' of knowledge production. It is for this reason that, even when there is controversy and contestation over particular issues, actors seek to frame themselves as relying on technoscience (as we saw in the context of activism around and appropriation of technoscientific knowledge, in Chapter 5, and the use of expert knowledge in contested policy issues, in Chapter 8). While technoscientific expertise is central to the functioning of many contemporary societies, we have also seen that it is perhaps more complex than it might appear (in particular in Chapter 7). Scholarship on the nature of expertise has framed it as multidimensional (incorporating more than simply access to knowledge) and as performed and relational. Expertise is spread throughout society, rather than being solely located within those with formal credentials, and it is also dependent on the ability of experts to convince others of their expert status. To think of expertise as something that can be more or less successfully performed calls attention to how it is negotiated in particular situations, and to the ways in which it can be mobilised by different actors: we saw, for example, how AIDS activists were able to develop 'cultural competences' that allowed them access to expert discussions, and how scientists themselves sometimes drew on other kinds of experiential knowledge to bolster their claims to expertise (in Chapters 5 and 7). *Expertise is therefore not a straightforward category*, nor one that we should take for granted or accept at face value. Rather than being a class into which certain types of people fit – those trained in science and technology, for instance – we should be attentive to how expertise is performed and contested in particular contexts, who is able to successfully present themselves as an expert, and on what grounds they do so.

A fourth idea is implicit to what I have already discussed. While representations and expertise do matter, it is clear that technoscience, its products, and the way in which it is represented in public spaces are never passively absorbed or received by non-scientists. *Publics actively engage with technoscience*, whether that is as engaged, tinkering users of particular technologies (Chapter 3), as active consumers of science and science communication (Chapter 5), as creators of technoscientific knowledge in research in the wild (Chapter 5), as activists who contest technoscientific issues (Chapters 3, 5, and 6), or as citizens participating in deliberation on scientific topics (Chapter 8). We should therefore anticipate that technoscientific knowledge and products will be encountered in the context of rich social worlds, and will be thoughtfully integrated into these. Even ignorance of science, we saw in Chapter 6, can serve particular purposes. It is thus vital not to write off protest or hesitation around technoscience as 'anti-scientific' in and of itself, or as something that would be straightforwardly solved by education. As in the notion of science-related populism (discussed in Chapters 1 and 7), resistance to or concerns about science are often tied to prior experiences (for instance of healthcare), long-standing political positions, and assessments of how trustworthy particular institutions are.

A fifth and related point is that *useful knowledge also resides outside of institutionalised technoscience*. In particular, we have seen that knowledge that is relevant to technoscientific research, choices, and trajectories is present in wider society as well as in academic spaces. This is captured through the notion of epistemic diversity, which we encountered in particular in Chapters 5 and 6, as well as in arguments for opening up science and technology through public participation and deliberation (Chapters 1 and 8). In Chapter 5 we saw that, while some activists aligned themselves with the methods and approaches of institutionalised technoscience, others used 'alternative epistemologies' to capture different aspects of environmental problems, or to work on timeframes different to those of mainstream science.[6] Similarly, indigenous scholars and activists have insisted on the integrity of indigenous and traditional knowledge (Chapter 5): such ways of knowing cannot simply be slotted into technoscientific paradigms, or used as an additional data source, but should be understood as knowledge systems that incorporate entire worldviews and sets of purposes and obligations that may differ from those of western science. It is thus vital to approach technoscience with humility. It is, as I wrote earlier, a hugely reliable system for creating knowledge about the world, but it is not the only source of useful knowledge, and indeed – as we saw in Chapter 6 in discussing unknown unknowns – it may not be well equipped to deal with some challenges.

This leads to a final idea, that *acknowledgement of and engagement with epistemic diversity is central to policy and to decision making on technoscientific issues.* Frameworks such as risk assessment simply do not capture the many questions that can and should be asked regarding the trajectories technoscience takes, and the directions in which it may take our societies (as discussed in Chapter 6). Knowledge and views from spaces outside of institutionalised technoscience are thus vital both to concrete decision making (hence, as we saw in Chapters 1 and 8, arguments for public participation) and to understanding any particular issue (such as experiences of disasters, as we saw in Chapter 6). The notion of epistemic diversity thus leads to that of epistemic justice, and to the question of whose knowledge counts, is valued and acted upon, in particular situations – an idea that I will return to later in the chapter.

The problem with 'innovation'

The central themes that I have just described start to point us towards questions of justice by raising the question of whose knowledge (whether technoscientific, experiential, or based on other grounds) is valued in particular situations. In the next sections I explore some more specific examples (and research about these) in order to illustrate how technoscience is entangled with power dynamics more generally, exploring spaces or processes where inequality and injustice play out through science and technology.

I start with an idea that has popped up in one or two places in the book so far: that of innovation. While this term is used flexibly (to say the least), it has become central to many policy agendas, framed as something to be promoted that will ensure economic growth and development.[7] 'Knowledge and innovation are the beating heart of European growth', said one 2005 European policy document.[8] This focus on innovation has been referred to as 'innovationism', or even a 'cult', and has been criticised as propagating a dangerous focus on novelty over and above the less hyped work of maintenance or repair (as discussed in Chapter 3).[9] We have met the term a number of times in the book so far: hackerspaces have been framed as sites of innovation led by laypeople (Chapter 5), for instance, while we also saw that efforts towards public participation and the 'responsibilisation' of science are increasingly oriented to bringing about socially robust and responsible innovation (Chapter 8). We have also encountered an implicit criticism of the kinds of policy language quoted earlier, in discussing the 'linear model' of research and technology development as a central myth that continues to circulate (Chapter 3). Innovation (and particularly its connection to basic research, on the one hand, and to economically generative technologies or products, on the other), is thus far more complex than promotion of it in policy might suggest.

This, then, might be one problem with innovation: that in practice it is rather hard to manage, direct, and bring about in any sustained manner.[10] Uncritical promotion of it is thus unlikely to have the desired effect, and may distract from other essential activities: as Lee Vinsel and Andrew Russell write, 'our obsession with the new undermines and devalues the mundane labor people do all around us, on which our lives depend each and every day'.[11] Another problem with a focus on innovation relates to what was discussed in Chapter 3 (in particular in the context of digital technologies): the way in which development of new tools and technologies tends to replicate a central pathology of technoscience more widely, that of imagining that research and development has a 'view from nowhere' – the idea that it is not located in any culture or identity but takes a 'god's eye' perspective on the world. By not acknowledging the situatedness of any form of knowledge production, and the ways in which it will always be shaped by specific human values and contexts, there is a danger that technoscience (and its products) work to universalise the experiences of some people while erasing those of others.

It is not new to argue that technoscientific knowledge production is in practice always situated: there is now a long tradition of feminist and decolonial scholarship (in particular) that has discussed how principles, ideas, or tools that are framed as universal are in fact oriented to the bodies and experiences of particular elites (often, White, high status, European men, in the context of the history of science). 'Objectivity' was something that had to be invented, and this invention was done from a particular position.[12] As a result, certain people have been able to present themselves as disembodied and disinterested observers of reality, while others, as Donna Haraway writes, are 'the embodied others, who are not allowed *not* to have a body'.[13] Those with bodies can only present partial perspectives, those who are 'unmarked' are able to confidently adjudicate on reality as a whole. Carol Dennis, writing in the context of the need to decolonise the university, notes that:

> The unmarked scholar requires no introduction. He does not need to explain his appearance in the text and he requires no further markers of qualification. What the unmarked scholar says is more important than who he is. He speaks from that place which is *just there*, that place which is no place. … This human status is not open to *all humans*. It is denied to females and those racialized as black.[14]

In practice, of course, there is no 'unmarked scholar', nor any 'view from nowhere'. 'We are entangled in our flesh, in our versions of vision, and in relations of power that pass through and are articulated by us. So detachment is impossible', writes John Law.[15] This has led to arguments of the urgency of situating knowledge claims (and technology development), and of acknowledging that any perspective is inevitably partial,[16] as well as contributing to discussions of the need to decolonise research practices by (among other things) situating (or 'provincialising') the work of authors from the Global North and others who continue to write as 'unmarked scholars'.[17] Rejecting the 'god trick' of imagining total objectivity would lead to humbler claims, but also to forms of technoscience that are more robust in that they reject 'unlocatable' and therefore 'irresponsible' knowledge claims, where '[i]rresponsible means unable to be called into account'.[18]

Box 9.1: Being Black in physics

'Who is allowed to be an observer in physics, and who is fundamentally denied the possibility?', asks Chanda Prescod-Weinstein, in an article titled 'Making Black Women Scientists under White Empiricism: The Racialization of Epistemology in Physics'. Prescod-Weinstein, a cosmologist and Black feminist theorist, analyses physics research practices to show that what she terms 'white empiricism' remains dominant: 'white men, who are the dominant demographic in physics, construct the figure of the observer to exclude anyone who does not share the attending social and intellectual identities and beliefs'.[19] This universalising of a particular subject position affects both the knowledge that is produced, and the status and recognition that is granted to particular physicists, making it harder for anyone outside of the norm of the White man – and particularly those at the intersection of multiple forms of oppression such as Black women – to gain prestige in the field. Importantly, physics is not alone in continuing to be structured around White supremacy and sexism. There continue to be racial disparities in successful applications for funding,[20] students of colour report invisibility or exclusion during their studies,[21] and gender bias affects the publication of research articles in many contexts.[22]

Perhaps these discussions seem abstract. At a time in which women's reports of pain continue to be disregarded as unreliable,[23] indigenous knowledge is ignored in curricula as not being knowledge at all,[24] and research practices continue to prioritise White perspectives,[25] however, they offer an important reminder that the myth that authoritative knowledge emerges solely from White male scholars continues to shape technoscientific practice, its entanglement with society, and individual experiences of it. And – to return to innovation – there is a very real risk that technology development will reproduce this 'view from nowhere' approach, developing tools and technologies that seek to be widely applicable (such as the Apple Watch) but that ignore certain bodies and identities (which in its first iteration did not include any form of menstrual tracking*). As we saw in Chapter 3, emerging digital technologies seem to be particularly susceptible to this problem. A now well-established lack of diversity in the tech industry and computer science more generally[26] means that limited perspectives, identities, and backgrounds are being mobilised when technologies – and their relation to the world around them – are imagined.† Indeed, developments in Silicon Valley through the 2020s such as the firing of critics or those working to diversify the industry[27] indicate that in many cases there is active hostility to the questioning of these dominant perspectives. 'Innovation' thus deserves critical interrogation with regard to whose needs, desires, and bodies are being assumed within it. In what ways might innovation – as a policy agenda and as it takes specific forms – be reproducing inequality, particularly with regard to whose views are prioritised and framed as universal?

Box 9.2: Innovation for who, by whom?

Smart cities are (according to the European Commission) cities in which 'traditional networks and services are made more efficient with the use of digital solutions for the benefit of ... inhabitants and business'.[28] They incorporate digital technologies into, for instance, waste management, traffic control, or lighting, and, as the involvement of the European Commission would suggest, are a focus for urban planning and policy in Europe and around the world. Bipashyee Ghosh and Saurabh Arora studied one smart city project in Kolkata, India, which ostensibly sought to use participatory, citizen-oriented methods to shape innovation. In practice, Ghosh and Arora found that only select citizen perspectives were heard (those belonging to the middle classes), and that the proposed innovative solutions were dominated by 'technology-centred "global" visions of smart urbanism'. 'To develop imaginaries that are genuinely responsive to the needs and knowledges of marginalised citizens', they write, 'modern technologies developed and sold by large corporations must be decentred, and grassroots innovations and practices must be foregrounded'.[29] Thus even in an innovation process that sought to be open to

* Of course, this omission should not cancel out the very real concerns women might have about sharing information about their reproductive health with mega-corporations, nor the privacy issues that are at stake in the use of such tracking tools. As we saw in Chapter 3, tracking by digital industry can itself become the focus for concern and (data) activism. Scholarship has thus explored the ways in which such tracking tools are used in practice.[b]

† Though it should be noted that there are some interesting counter-examples to Silicon Valley – for instance, the dominance of women in India's space technology sector. This also relates to what gets labelled as being 'tech', and celebrated as such: solely software development, or the other kinds of work (from marketing to the production of training datasets) that enable technologies to be brought into the world?

citizen participation, technocratic approaches dominated, and smart city plans became oriented to the visions and needs of the technology's developers and proponents. In this regard Kolkata is not unusual: at least some observers have suggested that smart cities are inherently undemocratic in that they promote 'neoliberal technocracy'.[30]

Other scholarship has questioned not only the forms that innovation is taking, and how these relate to justice and power relations, but the way in which the notion has come to dominate imaginations of social progress and of policy oriented to bringing this about. Innovation is consistently framed as a good thing, even though in practice it is not clear that it will uniformly lead either to economic development or to the public good (as discussed earlier and in Chapter 3). Critics have thus increasingly questioned the assumptions that are driving the promotion and celebration of innovation. At the very least, it is clear that 'disruptive' innovations such as social media or platform capitalism (the latter term refers to businesses such as Uber or microtasking platforms such as Amazon Mechanical Turk) have mixed impacts on society, impacts that it may have been useful to collectively reflect on before being rolled out in real-world experiments (as is the basic premise of concepts such as responsible research and innovation). Others have argued that to focus on innovation and economic growth is in and of itself questionable. How is it, asks Stevianna de Saille in the introduction to a book titled *Innovation beyond Growth*, that innovation:

> has become the be-all and end-all driver of economic growth, as if there were no other form of economic activity, no other way to produce what we need to live good lives? How much growth is really desirable, and of what kind, when ever larger sections of the population in historically richer nations are benefiting less and less from increases to GDP [Gross Domestic Product], while the environment and its biodiversity suffers more and more from the economic activity we already have?[31]

De Saille points to a set of central questions. Do societies really need to consistently grow economically, and what are the environmental costs of doing so? Is growth good when benefits from it are distributed unequally? Are monetary gains – assessed by metrics such as GDP – the only, or best, means of evaluating whether something is to the public good, and understanding the wellbeing of a society? De Saille and her colleagues therefore seek to disentangle technoscientific innovation from the idea that economic growth is the correct aim for any activity that seeks to serve society, questioning the way in which responsible technoscience is framed as needing to result in monetary value. Ultimately their argument is for 'responsible stagnation', the idea that in some contexts slowing down innovation may be the most responsible course of action. Others have similarly challenged an imagination of innovation as necessarily market- and technology-oriented. Writing in the context of development studies, where there is a similar reliance on a rhetoric of innovation as the central means for societies to industrialise or otherwise improve the wellbeing of their citizens, Andrea Jimenez and colleagues explore whether 'innovation in development [can] move from the Western-driven model of unsustainable growth and instead include alternative epistemological perspectives that are more sustainable', arguing that dominant notions of innovation 'reproduce a logic of inclusion that fails to avoid the pitfalls of neoliberalism'.[32] To frame 'knowledge and innovation' as the 'beating heart of European growth' (to quote again the policy document mentioned earlier) is thus to uncritically accept a logic that not only ties together technoscientific knowledge and technoscientific innovation, but that further assumes that all innovation is technoscientific, economic gains are the primary purpose of innovation, and that such gains will somehow will relate to public good. Presented in such terms, innovation is, if not necessarily a problem, at least something we might want to interrogate rather than assume to be straightforwardly positive.

Box 9.3: Techno-solutionism and humanitarian aid

There is a long history of optimism regarding the potential of new technologies to solve crises or meet basic needs in poorer countries. Tom Scott-Smith tells the story of the 'IKEA shelter' (properly called the 'Better Shelter'), a 'modular, sustainable, long lasting, recyclable, easily assembled, affordable, and scalable' shelter supported by the IKEA foundation and developed for what was framed as a 'refugee crisis' in Europe in the 2010s. The project won Design of the Year in 2017, received large amounts of press coverage, and was at least initially supported by the United Nations Refugee Agency. But it was also subject to increasing criticism: as well as fire safety issues that led to cancelled orders in Switzerland, Scott-Smith writes that the shelter 'was described as sustainable when in fact it involved flying piles of metal and plastic around the world [and] ignored established practice in the humanitarian shelter sector, which advocates the use of local materials and abundant local labor'.[33] Ultimately, media reporting had suggested 'that an intractable problem had been solved. It had not. Managing refugee arrivals is a complex political issue that requires sustained political engagement, legal reform, and advocacy in host states to ensure investment in welfare and protection'.[34] A single product, however sophisticated, could never meet these complex needs. Excitement about the Better Shelter thus in many ways aligns with what has been called, variously, techno-philanthropism, humanitarian neophilia, or simply techno-solutionism. Such approaches prioritise technological fixes developed in the west, focus on individualised solutions rather than systemic interventions, and often function as part of capitalist economies ('doing well by doing good'), as well as ignoring the experiences and preferences of those whose needs are being targeted.[35]

Towards a critical politics of technoscience

Innovation is thus one space in which we can start to explicitly engage with the politics of technoscience, and to observe how it becomes aligned with wider projects such as capitalism or neoliberalism through public discourse and funding practices. In this section I continue this engagement with the politics of technoscience by discussing two further specific entanglements: the way in which technoscience is and has been constituted through colonialism; and the nature of contemporary academic work. The latter, in particular, takes us away from spaces commonly labelled as 'public' or 'society' and into academic environments, practices, and interactions (though these are, as we saw in Chapter 8, shaped by public policy and funding decisions). We have thus far seen only limited aspects of such academic spaces because they are less overtly sites in which technoscience and wider society are co-constituted, but they remain key to how science is made in public by shaping the technoscientific resources that can move into more public contexts. While the ways in which scientific knowledge is made are not the focus of this book, we should at least briefly consider how the politics of this come to shape the processes and sites described in the volume so far.

What kinds of dynamics and trends are currently structuring academia around the world? While we have encountered arguments that technoscientific knowledge production is being opened up, or is increasingly oriented to the needs of wider society (in Chapters 2 and 8, in particular), other transformations are also at play. In discussing policy for science in Chapter 8 we covered developments such as the increased use of metrics and other forms of evaluation; such changes have been described as part of an academic 'audit culture' in which scholarly assessment is not only carried out by one's direct peers – other researchers who engage with and respond to academic work in publications and other forms of scholarly discussion and dissemination – but by university managers or those involved in other forms of research assessment (such as national research evaluations).[36] Simultaneously academic work is becoming *internationalised* (both with regard to expectations of international mobility by researchers and of scholarly dissemination at an international rather than only national or regional level);[37] *marketised* through the imposition of capitalistic logics of competition (in student recruitment and in research, as competition for funding or positions increases);[38] and *projectified* as university funding becomes increasingly oriented to short-term funding for research projects (leading to a concurrent casualisation of the academic workforce, with – in many research systems – the majority of scholars being employed on short-term, precarious contracts).[39] There is increasing concern, and public debate, regarding what these changes are doing both to the nature of university research and teaching careers, and to the knowledge that is produced. Perhaps such intense pressures are driving researchers to questionable research practices (discussed in Chapter 8), for instance, or are forcing others out of scientific careers altogether (for instance because of demands for international mobility despite caring responsibilities or a commitment to a particular place). What is clear is that in at least some research systems there is increasing unrest regarding academic working conditions: as I have been writing this book, there has been extended industrial action in universities in the UK, protests against a new university law limiting contract extensions in Austria,[40] and a new US book, *Organize the Lab*, which seeks to offer tools to help resist a situation in which:

> As someone who works in a lab, I am expected to work through the weekend and past my contracted hours in the week. I like the idea of science – it is exalted, useful, and interesting. Yet to be a scientist under a capitalist system I have to respect a hierarchy, publish or perish, and embrace long hours, low pay, and precarity that continues for years.[41]

Critiques of the labour conditions of academia cannot be separated from discussion of the kind of knowledge that is produced by exploited workers. It is already clear that many researchers (of necessity) orient themselves to the kind of scientific work that will be rewarded by contemporary systems of evaluation – for instance by focusing on 'high impact' publications rather than publishing negative findings, incremental work, that based on 'engaged' research based on collaborations with practitioners, or by writing in English rather than in other languages that might be more accessible to local research users. Similarly, it seems at least possible that the pressure to publish (the notion of 'publish or perish' mentioned in the quote), intense competition, and the expectation of long hours and unpaid labour is in some cases and contexts leading to low quality research. '[L]ong-term epistemic agendas', writes Max Fochler – referring to scientists' preferred scholarly programmes and interests – 'may be sacrificed to the needs of short-term productivity'.[42]

Box 9.4: The occupied university

There are increasing concerns regarding the conditions under which academics in many research systems work. Willem Halffman and Hans Radder, in their polemic 'Academic Manifesto: From an Occupied to a Public University', argue that the university has been occupied by management such that it no longer fulfils its public mission: 'the Dutch university', they write, 'is no longer there for the whole of society. No more science shops, no public university, no university as a platform for uplifting the people, but instead privatised knowledge embedded in expensive patents, published in unaffordable and exclusively English-language academic journals aimed at international colleagues and businesses'.[43] At the same time, western academic spaces remain highly exclusionary, occupied by and structured through ableist, cis-heteronormative, sexist, and racist norms. As an example, one study of disability in UK universities showed how hostile these environments were to academics with disabilities. 'Disabled academics', write Jennifer Remnant and colleagues, 'are subject to managerialist policies that position them as detrimental to the performance of the institution, rather than as organisational actors whose distinct experiences may enhance the work of a university'.[44] Academic work environments thus function to exclude those outside of an imagined norm, including people of colour, queer people, and those with disabilities. Simply existing in such spaces can take a toll (as is also documented in contemporary fictional accounts of life in science such as Brandon Taylor's *Real Life*[45]).

What do these developments mean for technoscience in public? While concerns regarding academic work are increasingly playing out in public media, we might also consider how the nature of contemporary academia may shape some of the sites and processes that we have encountered in the preceding chapters. Take the figure of the expert, for instance, and who is able to perform this role (as discussed in Chapter 7). If academic positions, credentials, and renown are useful to such performances, who is best able to achieve these based on the situation I have just described? Many of the evaluation processes I have mentioned prioritise or advantage particular kinds of research over others. The notion of 'excellence', for instance, which is often a key criteria within research evaluation, has been criticised as focusing on a particular kind of excellence: someone who is highly internationalised and mobile, prolific with regard to publications, and who carries out 'risky' or 'frontier' research. Critics have pointed out that not every researcher can achieve these characteristics, no matter how talented they are; those with families or care responsibilities, for example, find it more difficult to move to other countries or to find the time to write journal articles or research proposals.[46] Social factors thus shape who is able to become an 'academic expert', whether that is through the operation of structural inequalities or funders' choices regarding criteria for excellence (which could, of course, be defined differently, and other behaviours rewarded, from teaching to engagement with local communities).

The nature of contemporary academia can similarly be related to the 'undone' science discussed in Chapter 6, and to the questions of representation mentioned in Chapter 4. Funding priorities as well as career pressures will shape the kinds of research that is done, and therefore that which is not: increasingly, undone research includes that which requires a timeframe of more than three to five years (which does not fit into a project logic), that which speaks to local questions and needs rather than international research priorities, or that which, in working with user groups or stakeholders, must be carried out and disseminated in local languages. The qualities and behaviours that are rewarded in academic systems can therefore directly relate to the possibility or otherwise of scholarship that engages with the epistemic diversity discussed in Chapter 5, and that espouses the kind of humility that I suggested is one implication of recognising such diversity.[47] Similarly, we have already seen that public representations are not innocent depictions of a static reality concerning technoscience, but interact with its practice in important ways. Interestingly, for instance, there is evidence that public communication that tells 'success stories' regarding academic careers may lead to frustration on the part of researchers who encounter a very different situation as they progress within science.[48] If science is depicted as a fun, equitable space that offers a viable career path – and if experiences on the ground are rather different, as suggested by the quote from *Organize the Lab* – then it is not surprising that researchers become frustrated, and are increasingly leaving science altogether.[49]

For reasons of space, I will not continue parsing out the ways in which the conditions of contemporary research – and how they shape technoscientific knowledge – come to impact the public spaces and processes described in the preceding chapters. But I hope the central point is clear: the politics of academic work are relevant to how technoscience is negotiated in public. To be interested in questions of justice and equity in technoscience-society relations is thus also to be concerned with labour conditions in academia, and with who is able to succeed within it. This latter idea, relating to the kinds of bodies and identities that are able to feel 'at home' in academic spaces, brings us to a second vital aspect of the contemporary politics of technoscience: the ongoing challenge of decolonising technoscientific knowledge production (to the extent that doing so will be possible).

While it is impossible to do justice to debates around decolonising research here, one important starting point for these is the idea, discussed in Chapter 2, that technoscience is an intrinsically colonial project. Many of its directions and aspects – from social research to botany or medicine – emerged in service to or out of resources provided by European colonisation of other lands. Knowledge production frequently continues to replicate these practices of appropriation or exploitation. Concepts such as data or platform colonialism make reference to the ways in which digital technologies are created through the extraction of data resources from the Global South, continuing a long history of value extraction from colonised sites.[50] In addition, as Abeba Birhane has written, the dominance of corporate actors from the Global North means that not only is 'Western-developed AI unfit for African problems, the West's algorithmic invasion simultaneously impoverishes development of local products while also leaving the continent dependent on Western software and infrastructure'.[51] The ubiquity of mega-corporations such as Meta or the fact that digital technologies are being imported by companies from the Global North – however well-meaning the act of doing so may be – means that technologies are transferred to very different cultural contexts, with very different needs, with minimal reflection on the impacts of doing so (ignoring, as Birhane notes, vibrant but under-funded landscapes of local technoscientific activity). Research and development in the digital space is therefore repeating a long history of paternalism and White Saviourism, one that also continues to be seen in other forms of technoscience such as medical research and healthcare.[52]

Box 9.5: Consent and violence in extractive research

As noted at the start of this chapter, the history of science and technology includes many examples of unethical, violent, or exploitative research, in which value was extracted from individuals or groups without consent or benefit, or harm was done to them. One of the most notorious cases is the Tuskegee Study, which took place from 1932 to 1972 in the United States and in which syphilis was left untreated in Black men in order to observe its (terminal) effects. It remains, as James H. Jones writes, a 'symbol of research malfeasance in which virtually every principle underlying the ethical treatment of human subjects of research was violated'.[53] While there are now extensive, and often highly bureaucratic, frameworks that seek to ensure ethical research practice, some studies may still be conducted irresponsibly or cause harm. In 2024 one journal, *Molecular Genetics & Genomic Medicine*, retracted 18 articles it had previously published after concerns arose that the datasets they analysed had been gathered without consent. As noted in an article about the case, the papers:

> are all based on research that draws on DNA samples collected from populations in China. In several cases, the researchers used samples from populations deemed by experts and human rights campaigners to be vulnerable to exploitation and oppression in China, leading to concerns that they would not be able to freely consent to such samples being taken.[54]

The case thus resonates with continuing concerns regarding data ownership, consent, and who benefits from research at a moment in which analysis of large

datasets, whether biomedical or from digital media, is central to much research and technological development.

Similarly, colonial legacies continue to shape the global landscape of research. The extraction of resources, exploitation of peoples, and imposition of particular forms of social organisation that colonialism involved has meant that even 'post' colonial states are still dealing with the long-term effects of colonialism, and in particular that there are huge resource disparities between technoscience in the Global North and South. Outputs such as publications are lower in the Global South because of 'high publication costs, lack of institutional support, lack of external funding, bias, high teaching burden, and language issues',[55] while even where research focuses on Southern sites and contexts it is often led by researchers from the Global North, who then receive valuable indicators of esteem such as lead authorships.[56] In such collaborations, write Miller and colleagues, '[l]ocal researchers may be uncredited for their contributions, forgotten in the publication process, or given inadequate recognition as an author'.[57]

The impacts and logics of colonialism thus continue to shape how technoscientific research is carried out.* In addition there is ongoing rejection or suppression of the knowledges of colonised nations or of indigenous peoples, and the imposition of institutionalised technoscience from the Global North as being, as Sujatha Raman writes, 'the sole bedrock of epistemic and practical justification'.[58] This point thus relates to the ideas of epistemic diversity that we have already encountered (in Chapters 5 and 6), and to the need to reclaim and celebrate the value of non-technoscientific ways of knowing. Writing about science communication in the African context, Sesam and Ibiyemi observe that:

> [T]he systems of education and knowledge production that have crystallised in postcolonial societies over time are an amalgam of sorts in which traditional knowledge and communication systems feature little, if at all … what we are left with are 'inherited' and 'alien' systems that are not efficacious by virtue of their being disconnected from the pre-colonial histories and postcolonial realities of African societies. … In particular, higher education systems in Africa almost unilaterally promote northern scientific and educational paradigms and consequently facilitate the erasure of traditional ways of knowing.[59]

---

* And, indeed, continue to shape some central imaginations of what technoscience is for. Discussion of colonisation of the moon or of planets in our solar system and beyond indicates how deep logics of colonialism run: ideas of resource extraction, expansion, and claiming new territory or grounds are still resonant, and even dominant, in some public and policy contexts.[c]

Such traditional knowledges, Sesam and Ibiyemi argue, are not only integral aspects of local cultures and heritage (and thereby identity), but offer vital insights into contemporary challenges such as the climate crisis. Acknowledging epistemic diversity and moving towards epistemic justice – which involves not only such acknowledgement, but 'tak[ing] inequality seriously and seek[ing] to work against it'[60] – is thus not only necessary for reasons of equity, but in order to respond better to central challenges of the contemporary world. Rendering alternative epistemologies more visible offers an opportunity to act against 'epistemicide' – the destruction of knowledge systems – and to work towards 'cognitive justice', rejecting colonial ways of knowing and thinking in order to surface not only diverse means of making sense of the world, but possibilities for emancipation and social justice.[61]

Efforts to decolonise scholarship thus relate as much to the theories and methods that are used in research as to its global infrastructures and financing. This is visible, for example, in moves to decolonise not just academia generally, but universities specifically.[62] While these developments take a variety of forms – from acknowledging the racist histories of institutions to interrogating how Whiteness is reproduced within them – one important aspect involves engagement with curricula and canons. Whose work is visible, celebrated, and valued in university education? As with the politics of academic labour, this question relates, suggests William Jamal Richardson, to the notion of undone science (discussed in Chapter 6). In a global academic system still dominated by Eurocentrism, the 'categorical writing off of colonised peoples and their societies as knowledge producers ensures that, at least within Western-defined academic spaces, certain ideas always remain unthought'.[63] Incorporating previously unthought ideas into university curricula through the inclusion of a wider range of knowledge producers, and seeking to shape universities into spaces that engage with diverse epistemic resources, is thus another means through which technoscience might be at least partially disentangled from its colonial origins.

Box 9.6: Creating an indigenous university

What might it look like to create spaces in which different knowledge systems can converge and learn from each other? Devenir Universidad is one answer to this question. Grounded in the Inga community in Colombia, it starts from their territory and millennia of experience of this and focuses on 'the ways in which the indigenous communities can protect and transmit their knowledge. It creates partnerships, audiovisual media, cartographies and architectural design, and stages public events and exhibitions to support the indigenous-led initiative of creating a university'.[64] Devenir Universidad is thus an effort to foreground indigenous, place-based knowledges and to weave these together with knowledge from elsewhere, including western science.[65] The university is not a place in the sense of a set of buildings (there is no ivory tower), but at the same time, '[k]nowledge is viewed as embedded in the environment and knowing something means becoming part of this field of meaningful relations'.[66]

Technoscience and (epistemic) justice

While the critical politics of technoscience I have outlined here – involving engagement with the need to decolonise knowledge production, the conditions of academic work, and the nature of innovation – are at least in part different from older traditions of radical or critical engagement with science, they demonstrate a continuing need to interrogate not just technoscience's entanglements with society generally, but with power and injustice more specifically. In this respect we have in many ways come full circle, to the ideas and practices of the radical scientists discussed in Chapter 2. Born out of wider activist movements of the mid-20th century, these scientists were radical in the sense of being attentive to social justice and concerned about the ways in which science and technology intersected with this; for them, technoscience could be a force for good, but required reflection and direction in order to achieve this. After being shuttered for some decades, the journal of the radical science movement, *Science for the People*, was relaunched in 2019, a sign, perhaps, of a sense of renewed urgency in engaging with technoscience's connections to power. In it – and in other spaces of debate and critique – both scientists and those outside of technoscience continue to reflect on technoscience's ongoing connections to the military, the ways in which it is implicated in imperialist projects, how it is used to reinforce inequality, and other 'questions of capital, power, ideology, and democracy in science'.[67] As the editors of the revived journal write, '[m]odern science as we know it was born out of the Manhattan Project, when the state guided the market's invisible hands toward large research enterprises for clear military-geopolitical goals'.[68] *Science for the People* thus provides a space for reflection regarding how technoscience might be done differently, in ways less aligned with capitalistic, militaristic, or non-democratic goals.

More generally it is clear that movements such as the degrowth agenda or calls to decolonise universities speak to the need to acknowledge not only that technoscience is constantly entangled with human choices, values, histories, and meanings, but that these are, by their very nature, political. Radical science has focused on technoscience as a capitalistic project; increasingly, however, concerns relate to capitalism not in isolation but 'in relation to indigeneity, race, gender, racegender, coloniality, agriculture, war, and technologies of communication and surveillance', as Kelly Moore writes.[69] Technoscience, as we have seen repeatedly throughout this book, *does things* to the world, just as it is itself shaped by human choices. To engage with its relation to power and injustice is therefore to ask (as individuals and, importantly, as collectives of different kinds): what worlds do we wish to bring into being, and how might science and technology help – not hinder – us in doing so?

References

[a] See Liboiron, M., Ammendolia, J., Winsor, K., Zahara, A., Bradshaw, H., Melvin, J., et al (2017). Equity in Author Order: A Feminist Laboratory's Approach. *Catalyst: Feminism, Theory, Technoscience*, 3(2), 1–17.

[b] For instance, Algera, E. (2023). Knowing (with) the Body: Sensory Knowing in Contraceptive Self-tracking. *Sociology of Health & Illness*, 45(2), 242–258. Lupton, D. (2016). *The Quantified Self.* John Wiley & Sons. See also https://www.theatlantic.com/technology/archive/2014/12/how-self-tracking-apps-exclude-women/383673/

[c] See: Haskins, C. (2018, 14 August). The Racist Language of Space Exploration. *The Outline.* https://theoutline.com/post/5809/the-racist-language-of-space-exploration

[1] See Rasekoala, E. (ed) (2023). *Race and Sociocultural Inclusion in Science Communication: Innovation, Decolonisation, and Transformation.* Bristol University Press.

[2] Dawson, E. (2018). Reimagining Publics and (Non)Participation: Exploring Exclusion from Science Communication through the Experiences of Low-income, Minority Ethnic Groups. *Public Understanding of Science*, 27(7), 772–786.

[3] Orthia, L.A. and Roberson, T. (eds) (2023). *Queering Science Communication: Representations, Theory, and Practice.* Bristol University Press.

[4] Saini, A. (2017). *Inferior: How Science Got Women Wrong – and the New Research That's Rewriting the Story.* Beacon Press.

[5] Collins, H.M. and Pinch, T. (1998). *The Golem: What You Should Know about Science.* Cambridge University Press.

[6] Ottinger, G. (2010). Buckets of Resistance: Standards and the Effectiveness of Citizen Science. *Science, Technology & Human Values*, 35(2), 244–270. Ottinger, G. (2022a). Misunderstanding Citizen Science: Hermeneutic Ignorance in U.S. Environmental Regulation. *Science as Culture*, 31(4), 504–529. Ottinger, G. (2022b). Responsible Epistemic Innovation: How Combatting Epistemic Injustice Advances Responsible Innovation (and Vice Versa). *Journal of Responsible Innovation*, 10(1), 1–19. https://doi.org/10.1080/23299460.2022.2054306

[7] De Saille, S. (2015). Innovating Innovation Policy: The Emergence of 'Responsible Research and Innovation'. *Journal of Responsible Innovation*, 2(2), 152–168.

[8] COM(2005) 24 final. 2005. *Working Together for Growth and Jobs – a New Start for the Lisbon Strategy*. Brussels: Official Journal of the European Communities.

[9] Invernizzi, N., York, E., Dréano, C., Kaşdoğan, D., Kenner, A., Khandekar, A., Okune, A., et al (2023). Innovationism Across Transnational Landscapes. *Engaging Science, Technology, and Society*, 9(2), Article 2. https://doi.org/10.17351/ests2023.2503. Vinsel, L. and Russell, A.L. (2020). *The Innovation Delusion: How Our Obsession with the New Has Disrupted the Work That Matters Most*. Crown.

[10] See, for example, Kline, S.J. and Rosenberg, N. (1986). An Overview of Innovation, in National Research Council, *The Positive Sum Strategy: Harnessing Technology for Economic Growth*. The National Academies Press, pp 275–306, at p 275.

[11] Vinsel and Russell (2020), p 17.

[12] Daston, L. and Galison, P. (2007). *Objectivity*. Zone Books.

[13] Haraway, D. (1988). Situated Knowledges: The Science Question in Feminism and the Privilege of Partial Perspective. *Feminist Studies*, 14(3), 575–599, at p 575, original emphasis.

[14] Dennis, C.A. (2018). Decolonising Education: A Pedagogic Intervention, in Bhambra, G.K., Nişancioğlu, K. and Gebrial, D. (eds) *Decolonising the University*. Pluto Press, pp 190–207, at p 192, original emphasis.

[15] Law, J. (2004). *After Method: Mess in Social Science Research*. Routledge, p 68.

[16] Haraway (1988).

[17] Bhambra, G.K., Gebrial, D., and Nişancıoğlu, K. (eds) (2018). *Decolonising the University*. Pluto Press.

[18] Haraway (1988), p 583.

[19] Prescod-Weinstein, C. (2020). Making Black Women Scientists under White Empiricism: The Racialization of Epistemology in Physics. *Signs: Journal of Women in Culture and Society*, 45(2), 421–447, at p 422.

[20] Chen, C.Y., Kahanamoku, S.S., Tripati, A., Alegado, R.A., Morris, V.R., Andrade, K., et al (2022). Systemic Racial Disparities in Funding Rates at the National Science Foundation. *eLife*, 11, e83071. https://doi.org/10.7554/eLife.83071

[21] Leyva, L.A., McNeill, R.T., Balmer, B.R., Marshall, B.L., King, V.E., and Alley, Z.D. (2022). Black Queer Students' Counter-Stories of Invisibility in Undergraduate STEM as a White, Cisheteropatriarchal Space. *American Educational Research Journal*, 59(5), 863–904.

[22] Krawczyk, M. and Smyk, M. (2016). Author's Gender Affects Rating of Academic Articles: Evidence from an Incentivized, Deception-free Laboratory Experiment. *European Economic Review*, 90, 326–335.

[23] For instance, see Harvard Health Blog (2017, 9 October). Women and Pain: Disparities in Experience and Treatment. *Harvard Health Blog*. https://www.health.harvard.edu/blog/women-and-pain-disparities-in-experience-and-treatment-2017100912562

[24] Bhambra et al (2018), chapter 5.

[25] Prescod-Weinstein (2020).

[26] See discussion in Abbate, J. (2017). *Recoding Gender: Women's Changing Participation in Computing*. MIT Press.

[27] Schiffer, Z. (2021, 19 February). Google Fires Second AI Ethics Researcher Following Internal Investigation. *The Verge*. https://www.theverge.com/2021/2/19/22292011/google-second-ethical-ai-researcher-fired

[28] European Commission (nd). *Smart Cities*. European Union https://commission.europa.eu/eu-regional-and-urban-development/topics/cities-and-urban-development/city-initiatives/smart-cities_en

[29] Ghosh, B. and Arora, S. (2022). Smart as (Un)Democratic? The Making of a Smart City Imaginary in Kolkata, India. *Environment and Planning C: Politics and Space*, 40(1), 318–339, at p 334.

[30] Cardullo, P. and Kitchin, R. (2019). Smart Urbanism and Smart Citizenship: The Neoliberal Logic of 'Citizen-focused' Smart Cities in Europe. *Environment and Planning C: Politics and Space*, 37(5), 813–830.

[31] Saille, S. de and Medvecky, F. (2016). Innovation for a Steady State: A Case for Responsible Stagnation. *Economy and Society*, 45(1), 1–23. https://doi.org/10.1080/03085147.2016.1143727. Saille, S. de and Medvecky, F. (2020). *Responsibility Beyond Growth: A Case for Responsible Stagnation*. Policy Press, p 4.

[32] Jimenez, A., Delgado, D., Merino, R., and Argumedo, A. (2022). A Decolonial Approach to Innovation? Building Paths Towards Buen Vivir. *The Journal of Development Studies*, 58(9), 1633–1650, at pp 1634, 1636.

[33] Scott-Smith, T. (2017). A Slightly Better Shelter? *Limn*, 9. https://limn.it/articles/a-slightly-better-shelter/

[34] Scott-Smith (2017).

[35] Collier, S.J., Cross, J., Redfield, P., and Street, A. (2018). Preface: Little Development Devices / Humanitarian Goods. *Limn*, 9. https://limn.it/articles/precis-little-development-devices-humanitarian-goods/. Haven, J. and Boyd, D. (2020). Philanthropy's Techno-Solutionism Problem. *Data & Society Research Institute*. https://knightfoundation.org/philanthropys-techno-solutionism-problem/. Scott-Smith, T. (2016). Humanitarian Neophilia: The 'Innovation Turn' and its Implications. *Third World Quarterly*, 37(12), 2229–2251. Scott-Smith (2017).

[36] Shore, C. (2008). Audit Culture and Illiberal Governance: Universities and the Politics of Accountability. *Anthropological Theory*, 8(3), 278–298.

[37] Ackers, L. (2008). Internationalisation, Mobility and Metrics: A New Form of Indirect Discrimination? *Minerva*, 46(4), 411–435.

[38] Fochler, M. (2016). Variants of Epistemic Capitalism Knowledge Production and the Accumulation of Worth in Commercial Biotechnology

and the Academic Life Sciences. *Science, Technology & Human Values*, 41(5), 922–948. Slaughter, S. and Leslie, L.L. (1997). *Academic Capitalism: Politics, Policies, and the Entrepreneurial University*. Johns Hopkins University Press.

[39] Courtois, A. and O'Keefe, T. (2015). Precarity in the Ivory Cage: Neoliberalism and Casualisation of Work in the Irish Higher Education Sector. *Journal for Critical Education Policy Studies*, 13(1), 43–66. Ylijoki, O.-H. (2014). Conquered by Project Time? Conflicting Temporalities in University Research, in Gibbs, P., Ylijoki, O.-H., Guzmán-Valenzuela, C., and Barnett, R. (eds) *Universities in the Flux of Time*. Routledge, pp 108–121.

[40] See Woolston, C. (2022). The Scientific Workforce in 2022. *Nature*, 612(7941), 803–805.

[41] Science for the People (ed) (2022). *Organize the Lab: Theory and Practice*. People's Science Network, p 3.

[42] Fochler (2016), p 943.

[43] Halffman, W. and Radder, H. (2015). The Academic Manifesto: From an Occupied to a Public University. *Minerva*, 53(2), 165–187, at p 175.

[44] Remnant, J., Sang, K., Calvard, T., Richards, J., and Babajide, O. (2024). Exclusionary Logics: Constructing Disability and Disadvantaging Disabled Academics in the Neoliberal University. *Sociology*, 58(1), 23–44, at pp 34–35.

[45] Taylor, B. (2020). *Real Life*. Penguin.

[46] Ackers, L. (2008). Internationalisation, Mobility and Metrics: A New Form of Indirect Discrimination? *Minerva*, 46(4), 411–435.

[47] Jasanoff, S. (2003). Technologies of Humility: Citizen Participation In Governing Science. *Minerva*, 41, 223–244.

[48] Felt, U. and Fochler, M. (2013). What Science Stories Do: Rethinking the Multiple Consequences of Intensified Science Communication, in Baranger, P. and Schiele, B. (eds) *Science Communication Today: International Perspectives, Issues and Strategies*. CNRS Editions, pp 75–90.

[49] Woolston (2022).

[50] See, for instance, Echterhölter, A. (2021). Formative Encounters: Colonial Data Collection on Land and Law in German Micronesia. *Science in Context*, 34(4), 527–552. Couldry, N. and Mejias, U.A. (2023). The Decolonial Turn in Data and Technology Research: What Is at Stake and Where Is It Heading? *Information, Communication & Society*, 26(4), 786–802. Salih, M.A. (2023). Facebook's Platform Coloniality: At the Nexus of Political Economy, Nation-state's Internal Colonialism, and the Political Activism of the Marginalized. *New Media & Society*, 146144482211449.

[51] Birhane, A. (2020). Algorithmic Colonization of Africa. *SCRIPTed*, 17(2), 389–409, at p 389.

[52] See, for example, Crane, J.T. (2013). *Scrambling for Africa: AIDS, Expertise, and the Rise of American Global Health Science*. Cornell University Press.

[53] Jones, J.H. (2008). The Tuskegee Syphilis Experiment, in Emanuel, E.J., Grady, C.C., Crouch, R.A., Lie, R.K., Miller, F.G., and Wendler, D.D. (eds) *The Oxford Textbook of Clinical Research Ethics*. Oxford University Press, pp 86–96, at p 96.

[54] Hawkins, A. (2024, 15 February). Genetics Journal Retracts 18 Papers from China due to Human Rights Concerns. *The Guardian*. https://www.theguardian.com/world/2024/feb/15/china-retracts-papers-molecular-genetics-genomic-medicine

[55] Eckmann, P. and Bandrowski, A. (2023). PreprintMatch: A Tool for Preprint to Publication Detection Shows Global Inequities in Scientific Publication. *PLOS ONE*, 18(3), e0281659.

[56] Cash-Gibson, L., Rojas-Gualdrón, D.F., Pericàs, J.M., and Benach, J. (2018). Inequalities in Global Health Inequalities Research: A 50-year Bibliometric Analysis (1966–2015). *PLOS ONE*, *13*(1), e0191901. Miller, J., White, T.B., and Christie, A.P. (2023). Parachute Conservation: Investigating Trends in International Research. *Conservation Letters*, *16*(3), e12947.

[57] Miller et al (2023), p 1.

[58] Raman, S. (2023). Response to: 'Looking Back to Launch Forward: A Self-reflexive Approach to Decolonising Science Education and Communication in Africa'. Decoloniality Opens Up New Epistemic Vistas for Science Communication. *Journal of Science Communication*, 22(4), Y02.

[59] Sesan, T. and Ibiyemi, A. (2023). Looking Back to Launch Forward: A Self-reflexive Approach to Decolonising Science Education and Communication in Africa. *Journal of Science Communication*, 22(4), Y01.

[60] Milan, S. and Treré, E. (2019). Big Data from the South(s): Beyond Data Universalism. *Television & New Media*, 20(4), 319–335, at p 325.

[61] See Santos, B.S. (2015). *Epistemologies of the South: Justice Against Epistemicide*. Routledge.

[62] See Bhambra et al (2018).

[63] Richardson, W.J. (2018) Understanding Eurocentrism as a Structural Problem of Undone Science, in Bhambra, G.K., Gebrial, D., and Nişancıoğlu, K. (eds) *Decolonising the University*. Pluto Press, pp 231–247, at p 234.

[64] Devenir Universidad (nd). *Home*. https://deveniruniversidad.org/en/home

[65] Biemann, U. (2022). The Forest as a Field of Mind. *Quaderni Culturali IILA*, 4(4), Article 4. https://doi.org/10.36253/qciila-2059

[66] Devenir Universidad (nd). *University*. https://deveniruniversidad.org/en/university

[67] Science for the People (nd). *About*. https://magazine.scienceforthepeople.org/about

[68] Science for the People (nd). *Letter from the Editors*. https://magazine.scienceforthepeople.org/vol25-3-killing-in-the-name-of/letter-from-the-editors-3. See also the following for an account of the many ways in which

technoscience has been developed in and through the military: Smit, W.A. (2001). Science, Technology, and the Military, in Smelser, N.J. and Baltes, P.B. (eds) *International Encyclopedia of the Social & Behavioral Sciences*. Elsevier, pp 13698–13704.

[69] Moore, K. (2020). Capitalisms, Generative Projects and the New STS. *Science as Culture*, 30(1), 58–73, at p 67.

10

Conclusion: Resources for Life in a Technoscientific World

This book is, as I began by saying, a resource. It brings together and connects you to different ideas, scholars, and literatures that have something to say about the relationship between science, technology, and society. My aim was to collate, synthesise, and summarise, doing as much justice as possible to the complexities of the diverse spaces we have encountered while still capturing their range and sweep. I hope that you can use it as a starting point for engaging further with the debates and moments that most interest you.

While I want this book to be a resource, a set of tools for living in technoscientific societies, I have chosen not to give instructions or directions. It contains few practical tips and no lists of how to navigate the contemporary world. The resources I offer are concepts and ideas that help us think about technoscience and society, rather than concrete suggestions as to how to act on them.

In part this is because there are no easy instructions for responding to a view of the world in which technoscience is always social, and society is shaped by technoscience. This view may change our imaginations, the ways in which we think about the role and place of science in society, but given the diversity of the kinds of interactions and mutual shapings that we have encountered it is clear that there can be no single account of how to navigate these. Indeed, many of the spaces and debates I have described demand individual judgements and choices. I cannot tell you how to act in response to arguments for the need to decolonise the university, acknowledge epistemic diversity, properly engage with the emergence of public concerns about technoscience, or view expertise as flexible rather than static (for instance). These discussions require personal responses that may, however, lead to collective action (as in the case of the radical science movement, citizen activism around environmental harm, or efforts to ensure responsible technology development).

At the same time it seems too lazy to leave you without any concluding thoughts or suggestions. I want to offer three ideas that, to me, offer a summary of the resources – the tools to think technoscience in society with – that are outlined in this book. And, while I think that it is up to you to decide what to do with these resources as a whole, I will try here to be a little more directive than I have been so far.

First, *technoscience is a human project, and embeds human values.* It is constituted through social processes as well as engagement with the material world. This means, I believe, that we should be attentive to the question of which, and whose, values are being embedded within it. What choices have led to a particular aspect of technoscience being as it is, whether that is a technology or an area of research? What forms of life are modelled or prioritised in it? Who benefits from it, and whose desires are driving it? Asking such questions highlights the human-ness of technoscience, and the ways in which it progresses not through some kind of neutral, deterministic unfolding, but to serve particular ends.

Second, *technoscience is never inevitable, and can be resisted.* Exactly because it is comprised through social processes and values, other versions of it are possible – other values can be incorporated into it. Even where it is entangled with injustice, inequity, bias, or violence, there is scope to call this out and to reject its imposition on our lives. Technoscientific products and logics may seem all-encompassing, and our agency limited, but we have seen repeatedly in this book that they can be resisted, whether through data activism that calls out algorithmic bias, citizen science that mobilises and insists upon the value of alternative epistemologies, efforts to incorporate deliberative democratic decision-making into research, or moves to decolonise curricula. If technoscience is a human, social project, then we ourselves can be some of the humans who come to shape it.

Ultimately, then, *technoscience is contingent. It could be otherwise.** One thing I hope that this book has equipped you to do is to notice some of these contingencies: the way that expertise is performed, and who gets to do this, for example, or the kinds of promises that are made about emerging technologies and the futures that these create. By noticing these contingencies we can question (again, both individually and collectively) what other possibilities might exist. It is not inevitable that technoscience becomes increasingly embroiled in the search for 'innovation' and unending economic growth, that the nature of technologies are defined by mega-corporations rather than shared discussion of public good, or that it reproduces centuries-old patterns of colonial extractivism. Part of our responsibility is, I believe, to

* Again, this notion is taken from a long tradition of Science and Technology Studies scholarship.[a]

imagine how the world could be otherwise, and how technoscience might serve this.[1] To repeat a question asked in the preceding chapter: what worlds do you want to bring into being?

I wish you all the best in finding your answer to this question, and in navigating the technoscientific societies you move through, live in, and come to shape.

References

[a] See Bijker, W.E. and Law, J. (1994). *Shaping Technology/Building Society.* The MIT Press. Calvert, J. (2024). *A Place for Science and Technology Studies: Observation, Intervention, and Collaboration.* The MIT Press.

[1] Calvert (2024).

Index

T